Shrink Sleeve Technology

Other Labels & Labeling books:

ENCYCLOPEDIA OF LABEL TECHNOLOGY
Michael Fairley

THE HISTORY OF LABELS
Michael Fairley and Tony White

DIGITAL LABEL AND PACKAGE PRINTING
Michael Fairley

ENVIRONMENTAL PERFORMANCE AND SUSTAINABLE LABELING
Michael Fairley and Danielle Jerschefske

CONVENTIONAL LABEL PRINTING PROCESSES
John Morton and Robert Shimmin

LABEL DESIGN AND ORIGINATION
John Morton and Robert Shimmin

LABEL DISPENSING AND APPLICATION TECHNOLOGY
Michael Fairley

CODES AND CODING TECHNOLOGY
Michael Fairley

LABEL EMBELLISHMENTS AND SPECIAL APPLICATIONS
John Morton and Robert Shimmin

BRAND PROTECTION, SECURITY LABELING AND PACKAGING
Jeremy Plimmer

DIE-CUTTING AND TOOLING
Michael Fairley

MANAGEMENT INFORMATION SYSTEMS AND WORKFLOW AUTOMATION
Michael Fairley

SHRINK SLEEVE TECHNOLOGY
Michael Fairley and Séamus Lafferty

LABEL MARKETS AND APPLICATIONS
John Penhallow

For the latest list please visit: **www.labelsandlabeling.com**

Shrink Sleeve Techology

Edited by
Michael Fairley LCG, FIP3 and FIOM3
Séamus Lafferty Ph.D

Shrink Sleeve Technology
First edition published 2017 by:

Tarsus Exhibitions & Publishing Ltd

© 2017 Tarsus Exhibitions & Publishing Ltd

Printed by CreateSpace, an Amazon.com company.

ISBN 978-1-910507-14-8

Contents

While every care has been taken to ensure the information, charts, diagrams and illustrations in this publication are correct at the time of publishing it is possible that technology, specifications, markets and applications, or terminology may change at any time, or that the editor's or contributor's research or interpretation may not be regarded as the latest accepted guidance in some parts of the world of labels.

The publishers therefore cannot accept responsibility for any errors of interpretation or for any actions, decisions or practices that readers may take based on the publication content and would advise that the latest industry supplier specifications, standards, legislative requirements, performance guidelines, practices and methodology should always be sought before any investment or implementation is made.

Foreword

It has been a pleasure and an honor to collaborate with Mike Fairley on the creation of this book, a book that began as a master class on shrink sleeve label converting, which Mike and I worked together to develop. In the labeling world, Mike's stellar reputation is only surpassed by his expansive knowledge of the industry, and I am honored to have had the opportunity to work alongside both him and the other skilled contributing authors on this project.

My intrigue with the shrink sleeve labeling industry spans the better part of two decades. Early into my tenure in the industry, I was struck by a simple observation – the observation that shrink sleeve labels were perceived as both *new* and *old* at the same time! They were 'new' to the many converters who had just discovered shrink sleeves, fascinated by both their complexity and their potential. They were 'old' to the many converters who had produced shrink sleeve labels for decades, continuously improving their craft and refining their quality. My travels led me to parts of the globe where converters convinced themselves and others that the technology was a fad that would come and go, while those same travels exposed me to converters who produced market-leading designs for the products they decorated. It still fascinates me today how something old can at the same time be perceived as something new!

The science of shrink sleeves is not new at all. In fact, Japan introduced the world to the first shrink sleeve label over 50 years ago. Yet here we are in present times, and we see no shortage of innovation and dynamism in the shrink sleeve label format. It continues to evolve and progress at a rapid pace.

This book is presented to readers as an introduction to shrink sleeve labels and a fresh look at how to guide the shrink sleeve label production process from start to finish. If the objective in reading this book is to understand how to produce a shrink sleeve label, then the objective is falling short. Rather, the objective must be to understand how to produce the *perfect* shrink sleeve label. After all, a shrink sleeve label is complex to produce, requiring great care and attention to detail during each step of the production process. In spite of its complexities, this label merits every ounce of effort invested, based on its proven point-of-sale impact on the shelves of retailers worldwide – when it is done right. Therefore, the objective of producing anything less than perfection would be a disservice to the brand owners who place tremendous value on the impact this label has on their products' performance, as well as a disservice to the immense potential that this labeling technology offers to the marketplace.

The chapters in this book follow the logic dictated by the production process, and the detail within the chapters will outline the nuances that, with each step, allow for the attainment of shrink sleeve labeling perfection.

Chapter 1 sets the stage by tracing the origins and growth of shrink sleeve labels and providing a market context and an overview of the market potential of this label format.

Chapter 2 represents the first step in the production process, film selection, and it explains the importance of selecting a film that best pairs with the contour and shape characteristics of the container to which the film will be shrunk.

Chapter 3 takes readers through the pre-press process, illustrating the elements and methodologies of graphic design for shrink sleeve labels, and walks through the thought processes necessary for the conceptualization and execution of the graphic design that transforms a two-dimensional printed item into a three-dimensional work of art.

Chapter 4 delves into the printing process – laying ink on the film substrate – a process governed by the learnings from Chapter 2 relative to film selection and from Chapter 3 relative to graphic design elements.

Chapter 5 introduces some steps with which many readers may already be familiar, as well as some new steps specific to shrink sleeve converting. The traditional steps of slitting and sheeting may already be well known, but the shrink-sleeve-specific step of seaming – forming the film into a tube, or sleeve – will likely be an entirely new concept. Moreover, the traditional steps carry some exceptional demands for the shrink sleeve converting process that are not typically required when engaging in other forms of label converting.

Chapter 6 speaks to the final steps of the process as the converted shrink sleeve label is applied to a container and is subsequently shrunk to the contours of that container. This step is meaningful as it signifies the point where the finished sleeve morphs from the two-dimensional to the three-dimensional.

Chapter 7 brings home the concept of this label format requiring nothing short of perfection, and it outlines what the format looks like when various steps of the production process deviate from specifications. Each mistake represented in this chapter represents an opportunity – an opportunity to learn and an opportunity to improve, inching ever closer toward that elusive – yet attainable – *perfect* shrink sleeve label.

In summary, the shrink sleeve label is an enigma – it is new and it is old at the same time. It presents an incredible set of challenges full of complexity and full of beauty. Because of this, the knowledge we possess today will likely make us fall short tomorrow. It is therefore my expectation that the pages that follow will deliver a thorough understanding of the basics, and that as you move forward as a practitioner in the world of shrink sleeve labels, your knowledge will expand beyond the content of the chapters of this book.

After almost two decades in the shrink sleeve market, I am amazed with what I continue to learn every single day. It is my hope that your journey in this exciting and dynamic industry proves as fun and as rewarding as that which I have experienced!

Séamus Lafferty, Ph.D.
President
Accraply, Inc.

Preface

Look at the many different label technologies today and it soon becomes apparent that shrink sleeve labels are one of the fastest growing of all the many different product decoration technologies. Originating in Asia in the 1960s, it was not until the 1980s that the technology began to develop significantly in Europe and North America, and only in the mid-1990s that full length sleeving and the use of steam for sleeving started to occur. So, still a relatively new process.

Nevertheless, growth in the new millennium has been quite impressive, with sleeve technology becoming ever-more sophisticated, creating new applications and markets, adapting to the potential of narrow-web and digital printing for shorter runs, as well as multiple versions and variations, personalization, different types of application machinery, and much more.

There is also no doubting that brand owners and marketing teams have become very attracted to sleeve technology over the past ten or so years ⊠ and it's not difficult to see why. High quality, full body decoration, maximum 360-degree branding space, the potential to decorate complex and intricate shapes, the ability to add tamper evidence, and also offer abrasion resistance and waterproofing of the printed image.

Traditionally produced using mid-to-wide web flexo and gravure printing, sleeve labels are also now being increasingly produced using the narrow-web flexo and digital printing processes, and becoming integrated into a label converter's portfolio of label decoration solutions.

However, while sleeve label production might seem to be just another printing opportunity for existing label converters, it has to be understood that it is far more complex than just printing on a film. There are many different films with different degrees of shrink capability to understand, a knowledge of the image distortion process during origination and pre-press, as well as knowing the shrinkability and demanding performance requirements of inks, the demands of the seaming process, and understanding how printed sleeves will perform in shrink tunnels and during shrinkage.

It was with all these considerations in mind that this Shrink Sleeve Technology book was conceived. Its aim is to educate and inform both new and existing sleeve label producers, to provide guidance on the film, ink, printing and processing stages, and to help understand why things may go wrong, how to recognize faults, and what remedial action may be required.

It is hoped that readers will find the book of value in the future growth of their businesses, in enhancing their quality and performance standards, in talking with existing and potential new customers, and in their general day-today sleeve production and problem solving activities.

Michael Fairley
Director, Labels & Labelling Consultancy
Founder, Label Academy

About the Label Academy

This book is part of the recommended study material for the Label Academy, a global training and certification program for the label industry. The Label Academy was created by the team behind Labels & Labeling magazine and the Labelexpo series of events.

The Academy consists of a series of self-study modules, combining free access to relevant articles and videos with paid text books (both printed and electronic). Once a student has completed a module, there is an opportunity to take an online test and earn a certificate.

It is expected that a Label Academy qualification will become a standard in the industry – for printers/converters, suppliers, brand owners and designers – and assist in providing a benchmark. In addition to its own training, the Label Academy will aim to become a resource provider to the many existing educational programs in the industry. Accredited training courses will be promoted through the Label Academy website and books will be provided at discounted rates.

The Label Academy concept was pioneered by industry expert Mike Fairley. This was in response to a reduction in the number of dedicated printing colleges and the need to standardize training across the world. The label industry also has its own specific training needs – it has some of the widest range of materials, printing processes and finishing solutions of any printing sector.

We are also working with other training experts and authors to ensure that the Label Academy provides up-to-date and relevant training material for the industry.

The Label Academy is supported by the key trade associations, including FINAT, TLMI and the LMAI.

www.label-academy.com

Label Academy sponsors

Thank you to our founding sponsors, without whom this ambitious project would not have been possible:

Cerm

Cerm designs business automation software solutions to meet the specific demands of flexo and digital narrow web printers. Using the latest technology, our team's focus is on innovation and continuous improvement.

Our automation solutions support each step in the printer's integrated workflow – from estimating to production, shipment and data collection – and provide the feature and functionality printers need to gain efficiency and improve profitability.

Cerm inspires collaboration and helps printers remain competitive in the market and deliver the best products possible. We are proud to sponsor the Label Academy and contribute to the future of the narrow web printing industry.

www.cerm.net

Flint Group Narrow Web

Flint Group Narrow Web has the products, the solutions, and the technical experts to handle any print situation. Providing solutions for food packaging, sustainability, increased bottom line, efficiency, and uptime – delivering the basics needed to run a successful operation, and the expertise to go above and beyond to another level of success.

Our experts provide solutions to your printing problems with the innovative products and services that have made us an industry leader around the world. Wherever you are, we are – available to help you reach your business goals today and into the future.

Continuous improvement is paramount to Flint Group; we are proud to sponsor the Label Academy and the benefits it will bring to the future of our industry.

www.flintgrp.com

Gallus Group

The Gallus Group with its production sites in Switzerland and Germany is a leader in the development, production and sale of narrow-web, reel-fed presses designed for label manufacturers. The machine portfolio is augmented by a broad range of screen printing plates (Gallus Screeny), globally decentralized service operations, and a broad offering of printing accessories and replacement parts. The comprehensive portfolio also includes consulting services provided by label experts in all relevant printing and process engineering tasks. The Gallus Group is a member of the Heidelberg Group and employs around 430 people, of whom 253 are based in Switzerland. The group headquarters is in St.Gallen, Switzerland.
www.gallus-group.com

MPS Systems B.V.

Producing high-quality label printing depends on several factors; one of them is the operator of the press.

As a press machine builder since 1996, MPS Systems B.V. knows how important training and education on subjects like pre-press, label printing and finishing is. For label printers, it is critical that their operators keep up with pre-press and press developments in addition to label trends. Therefore, MPS sponsors the Label Academy, to advance operator's passion for printing, share expertise and help multiply benefits.

The MPS slogans of 'Printers First' and 'Technology with Respect' have always underlined the core philosophy of MPS from press design to operator satisfaction. We develop our presses with a strong focus on user-friendliness and respect for the press operator: Printers First.
www.mps4u.com

HP Indigo

HP Indigo is a global leader in digital printing, with a broad portfolio of digital presses and workflow solutions. Indigo's proprietary Liquid Electrophotography (LEP) technology delivers exceptional print quality for the widest variety of applications including labels, flexible packaging, shrink sleeves and folding cartons. HP Indigo's digital presses match gravure print quality satisfying the most demanding brands.

A division of HP Inc.'s Graphics Solutions Business, Indigo serves customers in more than 122 countries, including many of the top label and packaging converters worldwide.
www.hp.com/go/labelsandpackaging

UPM Raflatac

In a little more than three decades, UPM Raflatac has become one of the world's leading manufacturers of pressure sensitive label materials, developing and leveraging the latest innovations in adhesive technology. Our film and paper label stocks are used for product and information labeling across a wide range of end-uses – from pharmaceuticals and security to food and beverage applications.

We are an engineering driven company with industry-leading products known for their consistent high quality and top performance. We are also known for the high performing supply chain and undisputed leadership in the area of sustainability. UPM Raflatac's dedication to innovation, sustainability and top quality is matched only by our commitment to service excellence. We call it the Raflatouch.
www.upmraflatac.com

Contributors

This Label Academy book on Shrink Sleeve Technology has evolved from the speaker presentations given at the five-hour Shrink Sleeve Workshop held at Labelexpo Americas 2016. Each speaker session was recorded, transcribed, added to in some cases, edited and approved by the respective speaker companies for publishing. The speaker presentations/editorial contributions used in this book are as follows:

Chapter 1
Michael Fairley
Managing Director
Labels & Labelling Consultancy

Chapter 2
Phil Heyworth
Group Vice President, Marketing and Business Development
Klöckner Pentaplast

Chapter 3
Bart Meersschaert
Senior Application Sales Manager
Esko

Chapter 4
Tom Hammer
Director, Product Portfolio
Flint Group Narrow Web NA

Chapter 5
Ben Ritter
Sales Executive and Sleeve and Converting Specialist
Accraply, Inc.

Chapter 6
Richard Howlett
Market and Product Line Leader, Shrink Sleeving
Accraply, Inc.

Chapter 7
Séamus Lafferty
President
Accraply, Inc.

About the Label Academy Founder

Michael Fairley
Director, Labels & Labelling Consultancy
Founder, Label Academy

Michael Fairley has been writing and speaking about label and packaging materials, technology and applications since the 1970s, both as the founder of Labels & Labeling and other print industry magazine titles and as an international consultant writing or contributing to label industry market and technology research reports for the likes of Frost & Sullivan, Economist Intelligence Unit, Pira, InfoTrends and Labels & Labelling Consultancy.

He is the author of the Encyclopedia of Label Technology, co-author of the Encylopedia of Brand Protection, a contributing author to the Encyclopedia of Packaging Technology and a contributing author to the Encylopedia of Occupational Health and Safety. He also provided significant input to the Academic American Encyclopedia, as well as writing more than a dozen technical books relating to paper, printing, labels and packaging, and pioneering the concept of the Label Academy.

He now works as a consultant to Tarsus Exhibitions & Publishing – which organizes the Labelexpo shows, Label Summits and publishes Labels & Labeling magazine – as well as regularly speaking at industry conferences and seminars.

He is a Fellow of the Institute of Packaging / Packaging Society, Fellow of IP3 (formerly the Institute of Printing), a Freeman of the Worshipful Company of Stationers, an Honorary Life Member of FINAT and a Licentiate of the City & Guilds of London Institute. He was awarded the R. Stanton Avery Lifetime Achievement Award in 2009.

Acknowledgements

The manufacturing and production stages involved in creating attractive and successful shrink sleeves are possibly more complex than any other method of product decoration. The filmic materials used possess specific properties, origination and pre-press require an image distortion process, inks need special shrinkable and performance properties, the shrink process requires an understanding of sleeve formation and seaming, shrinking, heat tunnels, and of possible problems and issues that may occur.

With this in mind it was determined that the industry's leading shrink sleeve materials and production experts should make up the panel of contributors. Speakers at the 2016 Shrink Sleeve Workshop alongside Labelexpo Americas were therefore approached and accepted that their sessions could be transcribed, edited and formatted to provide an industry-leading education handbook.

These book contributors and their affiliations are listed on the 'Contributors' page. Special thanks must be given to all of them, not only for their original speaking contributions, but also for their subsequent input of time and effort that went in to reading, amending, updating and agreeing their individual chapter contents. Their valuable input into this comprehensive guide to shrink sleeve technology is much appreciated.

Additional thanks are due to Jan De Roeck, Director Solutions Management, Esko, for editorial input to the chapter on design, graphics, distortion, visualization and origination for sleeve labels.

Finally, a special and heartfelt thank you to Séamus Lafferty, President, Accraply, Inc., and to Sean Murphy, Accraply Market Manager for Latin America, for the many hours they spent reading through the various book chapters, adding to or editing content, re-writing elements, making suggestions and providing additional illustrative material. Without their tireless input this book would surely not be as comprehensive or detailed as it now reads.

Additionally, Séamus Lafferty kindly agreed to write the Foreword for the book.

The future development and growth of shrink sleeve technology will owe much to Mr. Lafferty, to Accraply, and to the other industry expert speakers and contributors to this valuable Label Academy book.

Chapter 1

An introduction to sleeves and sleeve technology

Shrink sleeve technology is one of the fastest growing of all the label and product decoration processes, and increasing numbers of converters are making the leap to begin producing shrink sleeve labels. But what is shrink sleeve labeling? When did it start? What does it offer that other labeling technologies cannot offer? What are some of the key markets and applications for which it is frequently used? What is the size of the shrink sleeve market, and at what rate is it growing?

This book will serve as an introduction to the art and science of the shrink sleeve label converting process. It is our hope to both educate and to provide the tools needed to either enhance an existing shrink sleeve output to a higher level of quality and consistency or, for converters contemplating entering the market, make an informed decision whether to make the leap.

To begin, what is sleeve labeling?

The broader category of sleeve labeling is comprised of three types of sleeves. The first is **stretch sleeves**, which are LDPE films that are formed into a tube and then stretched and pulled over the container (Figure 1.1). The stretch in the film is what allows it to form a tight fit onto the container. Stretch sleeves comprise a very small portion of the sleeve label market, and these sleeves are typically found on containers with few contours or shapes.

The second sleeve labeling category is generically known as **machine direction orientated** (MDO) sleeves and encompasses both R.O.S.O.™, or roll-on shrink-on, and roll-fed-shrink or roll-applied-shrink technologies. This roll-fed or MDO shrink label material is applied in one of two ways: R.O.S.O.™ is

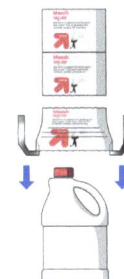

Figure 1.1 The process of stretch sleeve labeling © 2017 Accraply, Inc.

applied directly to the container in the same fashion as a traditional wrap-around (or roll-fed) label would be applied (see Figure 1.2). It is unwound and wrapped around the container, joined with an adhesive, solvent, or ultrasonic or laser weld. Following application, the product travels through a heat tunnel to shrink the label to the contours of the container. Applied in the same way as conventional wrap-around labels, R.O.S.O.™ labels offer full coverage and up to 15% - 18% shrinkage on

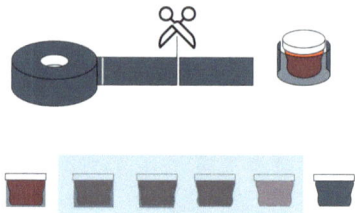

Figure 1.2 The process of roll-on shrink-on labeling © 2017 Accraply, Inc.

a contoured container.

Roll-fed shrink (RFS) or roll-applied shrink are terms typically associated with MDO films that have higher levels of shrink, and are therefore able to be applied to containers with more elaborate shapes and contours.

Because of the more elaborate shapes, the containers are less suitable to act as the mandrel around which the film is wrapped in the application process. For this reason, the application process becomes more complex and requires that the sleeve material first be wrapped around a mandrel that is sized appropriately, based on the dimensions of the container that will be sleeved. It is on this mandrel that the material is welded before being dropped over the container. This more sophisticated approach is illustrated in Figure 1.3.

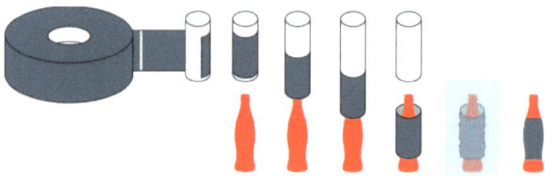

Figure 1.3 Roll-fed shrink film being applied to a mandrel prior to being dropped over the container © 2017 Accraply, Inc.

In summary, roll-on shrink-on labels are an ideal combination of wraparound and heat shrink sleeves that can be used to decorate a variety of container types such as plastic, glass and metal. Moreover, it is an ideal process for high speed labeling. Individual yogurt drinks, aerosols and glass bottles are just a sampling of some of the many products that are sleeved using this type of label technology.

The third sleeve technology is **heat shrink sleeve labeling**, which will be the focus of this book. The heat shrink sleeve labeling process involves selecting the appropriate shrink sleeve film, printing on the film, forming it into a sleeve with a welded seam and then cutting the sleeve tube to the required length, dropping it over the container and then shrinking it through a heat tunnel.

Shrink sleeve labeling remains the dominant sleeve technology globally. It comprises over 80% of the sleeve market and continues to grow. It remains one of the fastest growing of all the labeling technologies.

Foundationally, the shrink sleeve labeling process centers on the heat shrinkability of specific films (which will be discussed in detail in this book), with the printing process typically taking place on the inside of the film rather than the outside. The printed label web is formed and sealed into a sleeve with a seam, and it is then collapsed and rewound. The seamed sleeve is cut to the required container length and applied over the container, either manually or automatically, with the sleeved container then going through a heated shrink tunnel. We illustrate this multi-step process in Figure 1.4.

As can be seen, the shrink sleeve process starts with the container at the top left-hand side of the figure. We select the type of film we will use, and we then proceed to the pre-press operations. Because the printed image needs to be shrunk in the heat tunnel, the pre-press image must be pre-distorted, which we accomplish using pre-press software, so that the final decorated container image ends up correct. The pre-press technology we use to accomplish this is sophisticated, and it is central to achieving an optimal three-dimensional image on the shrink sleeved container.

As we move along the diagram to the printing process, we must be very mindful of the inks we

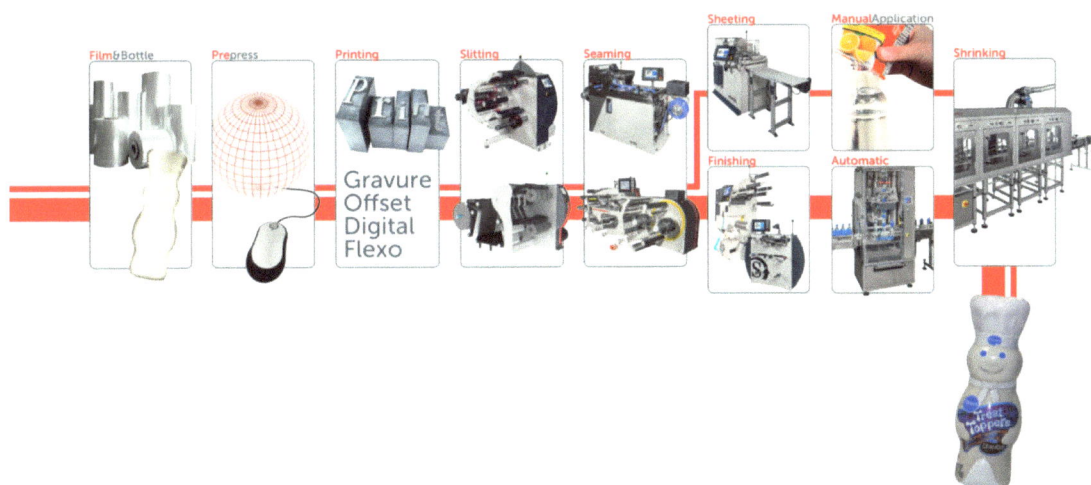

Figure 1.4 The complete shrink sleeve process, step by step © 2017 Accraply, Inc.

select to accommodate for the heat that we use to shrink the sleeve. The printed sleeve web is then slit, slightly oversized, into the individual container circumference widths. From the slitting process, the slit, printed web is formed into a sleeve, or tube, with an overlapped joint on a seaming machine. Finally, the sleeved film is collapsed, inspected, and rewound for shipping to the sleeve decoration plant.

Once the sleeved film arrives at the sleeve decoration plant, we will unwind the film back to a tube, cutting the sleeves to the appropriate container depth, and then the sleeve will be dropped over the container using manual or automatic application. Once we have placed that sleeve on the container, it will pass through a shrink tunnel, and upon exiting the tunnel, we will have our finished product – a shrink sleeved container. We will examine each of these stages of the heat shrink sleeve labeling process in detail in this book.

THE INTRODUCTION AND EARLY GROWTH OF SHRINK SLEEVE LABELING

Where did shrink sleeve label technology start? Shrink sleeve labeling hails from Japan, more specifically from the Fujio Carpentry Shop. Introduced in the 1960s, the earliest shrink sleeve example was

SHRINK SLEEVE TIMELINE

- 1965 - First use of shrink sleeves as labels by Fujio Carpentry Shop
- 1967 - Fujio Carpentry Shop changes its name to Fuji Seal Company
- 1970s - Shrink sleeves started to be used in Europe for promotional twin packs
- 1980s - Japan introduces single product shrink sleeves to Europe and North America
- Mid 1980s - Large-scale entry of shrink sleeves into the packaging market
- 1995 - First full length sleeving of narrow neck bottles
- 1995 - First application of translucent full length shrink sleeves
- 1996 - First use of steam for sleeving
- 2003 - First ever sleeving of Coca-Cola glass bottles
- 2006 - First introduction of recyclable shrink sleeves for use on PET containers

Figure 1.5 Evolution of shrink sleeve technology. Source: Sia Consulting

functional in nature: it was a means of providing tamper evidence when sake changed its packaging from wooden barrels to glass bottles. This change led to the introduction of the first tamper evidence seals using PVC film. From the early example that the Fujio Carpentry shop – now Fuji Seal Company – produced for sake bottles, we have seen the shrink sleeve industry evolve and change with the demands of the market (Figure 1.5).

Figure 1.6 Some of the many different shaped bottles that can be decorated with shrink sleeves © 2017 Accraply, Inc.

The 1970s marks when shrink sleeve labeling expanded outside of Japan to parts of Europe, but it was not until the 1980s that shrink sleeve labels saw rapid growth in Europe, followed by an entry into the North American market as well. By 1995, the shrink sleeve labeling industry realized significant advances, including the ability to apply shrink sleeve labels to full-length narrow neck bottles and containers of different sizes and shapes. Just one year later in 1996, the industry saw yet another advance – the use of steam as a heat source for shrink sleeve finishing. Greater concerns about preserving the environment led to the development of the first recyclable shrink sleeve label in 2006, yet another example of the industry innovating to meet the evolving needs of the market.

In recent years, we have observed the expansion of shrink sleeve labeling beyond just high volume applications. Digital pre-press has brought greater levels of sophistication and technology to shrink sleeving, and as a result, we have seen narrower mid-web presses – rather than just wide-web presses – getting in on the action. Furthermore, there have been improvements in flexo, UV inks, LED UV inks and digital printing technologies. With these advances, converters are now able to ask the question, "How can we produce sleeves on the technology that we're using now?"

The future for the shrink sleeve industry appears especially bright. We are seeing advances with narrow- to mid-web presses for short runs, and we are seeing advances in application machinery and more efficient lower speed machinery, all of which contribute to further growth in this exciting industry.

SHRINK SLEEVING TODAY

The advances in the shrink sleeve market have taken product decoration to an entirely new level. What we see in the market today is high quality, full body container decoration that maximizes branding real estate on the container. It offers the decoration of complex and intricate shapes (Figure 1.6). Additionally, shrink sleeve labeling has expanded to cans, providing smaller beverage customers with a more cost-effective option to ordering vast quantities of

Figure 1.7 Full body shrink sleeve decoration on cans © 2017 Accraply, Inc.

printed cans (Figure 1.7). Other advances include the incorporation of tamper evidence into the sleeve and providing hidden coding inside the sleeve as well. Finally, contemporary shrink sleeve labels may allow for reduced wall thickness of plastic and glass bottles and increase container strength and rigidity. Generally speaking, shrink sleeve films are durable, abrasion resistant and waterproof – the image is printed on the inside of the sleeve, as mentioned – and therefore make shrink sleeve labels a more and more compelling product decoration solution today.

SHRINK SLEEVE PRESENCE IN THE MARKETPLACE

We need not go far to encounter shrink sleeve labels in the marketplace today. Figure 1.8 shows the key categories where we see shrink sleeve labels in order of market size, and it is no surprise to see that beverage, dairy and food are the key industries where shrink sleeves are most widely used. What we observe as markets advance in their adoption of shrink sleeve labels, however, is that brand owners in categories such as detergents, cosmetics, pet food, paints and other consumer goods begin to adopt these labels as well. It is fascinating to observe the breadth of container profiles that shrink sleeve labels decorate in the marketplace: containers with narrow necks, concave shapes, convex shapes, and even bottles with multiple contours. Shrink sleeve labels have unleashed new levels of creativity, and the potential for uniquely shaped containers paired with eye-catching designs is almost limitless.

SLEEVE LABELS BY APPLICATION

1. Beverage - Energy drinks, juices, spirits, beers
2. Food - Dairy products
3. Toiletries, Health and Beauty
4. Household cleaning products - Detergents, soaps, cleaning agents
5. Pharmaceutical and Neutraceutical - Product safety and tamper protection
6. Packaged consumer goods/Retail

Figure 1.8 Main industry applications for shrink sleeve labels

HOW SIZABLE IS THE SLEEVE LABEL MARKET?

To provide some context for how significant in size the sleeve market is relative to other types of labels, see Figure 1.9. As noted in the graphic, the sleeve label market is steadily approaching one-fifth of the total world label market for all types of labels.

Self-adhesive (or pressure-sensitive) and wet glue labels are the largest global label types by volume at 36 and 39 percent, respectively. However, when the Westernized markets are examined more closely, we find that self-adhesive labeling is around 42 to 43 percent of the market, with wet glue comprising some 32 to 35 percent. Shrink sleeve comprises a greater percentage of the total market when we examine specific geographies. For instance, Asia, and more specifically Japan, is the largest market for sleeving as it invented the technology and is to this day at the farthest point of the adoption curve.

The annual growth forecast for sleeve technologies is currently 4.5-6.0 percent per annum in the latest

study that AWA published in mid-2016. A report published in December 2016 by Persistent Market Research (PMR) indicates that the sleeve market is showing annual gains of 5.2 percent. The growth for all types of labels is around 3.5 percent worldwide. In fact, the best data available pertaining to the growth of the labeling industry shows an annual growth rate

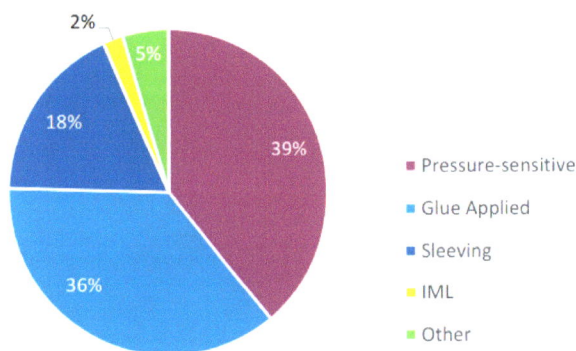

Figure 1.9 World market shares by labeling technology
Source: AWA 2016 Global Sleeve Label Market Study

of between 3.5 and 6.0 percent every year for the past 30 years - and forecasts tell us that we can expect this trend to continue for years to come.

Taking a deeper dive now into sleeve label data, we begin with an approximation of the size of the total market, which is estimated by AWA to be 10,500 million square metres of label stock. Shrink sleeve technology dominates the total sleeve labeling market at 87 percent of the market, stretch sleeve comprises nine percent, and R.O.S.O.™ comprises the balance. The annual growth forecast for the sleeve technologies is somewhere between 4.5 and 6.0 percent per annum presently, making sleeving the fastest growing of all the labeling technologies

GLOBAL GROWTH PATTERNS AND TRENDS

Where are the major global markets for sleeve label technology? We already know that it was invented in Japan, and as we observe in Figure 1.10, the Asia Pacific market continues to be the dominant user of this technology by a significant margin. This region is also estimated to register the highest CAGR, largely attributed to the mounting demand for beverage and packaged food in China, India and ASEAN countries.

As was mentioned earlier in this chapter, shrink sleeves entered the European market shortly after penetrating the Asian market, and Europe accounts for some 20% of the world label market today, though

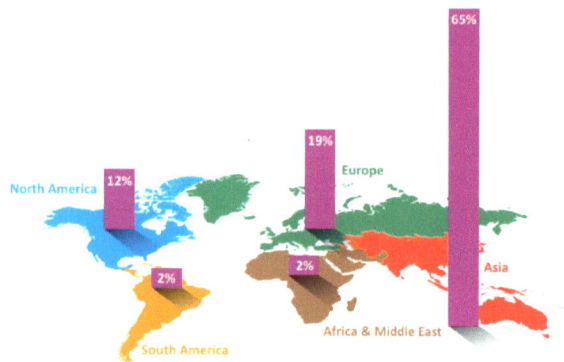

Figure 1.10 The global sleeve market. Source: AWA 2016 Global Sleeve Label Market Study

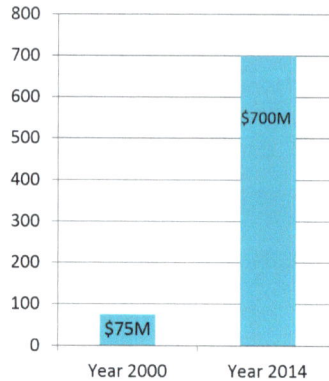

Figure 1.11 The shrink sleeve market in the USA. Source: industry estimates

European sleeve labeling growth is forecast to be somewhat uneven over the coming years.

North America followed Europe's lead with shrink sleeves several years after the technology took hold there, but the diffusion and growth in North America happened at a consistent rate. Still a slightly smaller market than Europe, North America now has multiple

Historical and forecast growth of sleeve label market to 2020

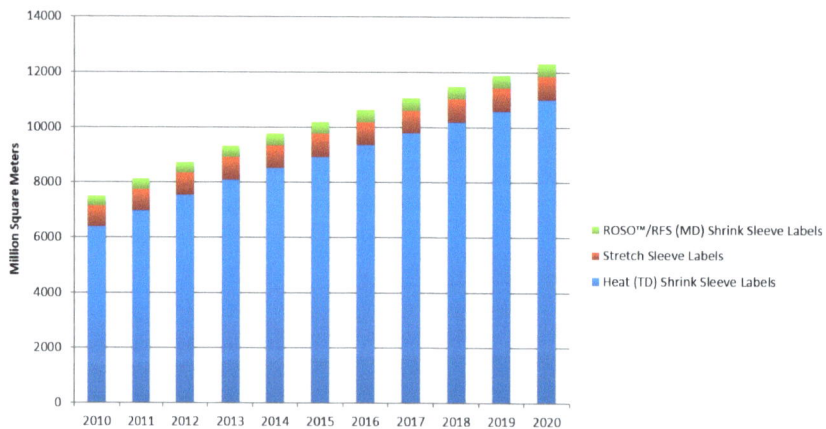

Figure 1.12 The historical and forecast growth of the sleeve label market to 2020. Source: AWA Global Sleeve Label Market & Technology Review 2016

SLEEVE LABEL RAW MATERIALS
- PVC – Polyvinyl Chloride
- PETG – Polyethylene Terephthalate Glycol-modified
- OPS – Orientated Polystyrene
- OPP – Oriented Polypropylene
- PE – Polyethylene
- PLA – Polylactic Acid

Figure 1.13 Films used for heat shrink labeling

KEY STAGES IN SLEEVE PRODUCTION
1. Origination and pre-press
2. Web-fed printing of film
3. Slitting web to label width
4. Forming slit web into a tube and seaming
5. Re-winding of seamed tube
6. Cutting tube to required depth and then applying to product container
7. Heat shrinking of film on sleeved product or container to a tight, shaped fit

Figure 1.14 Key stages in sleeve production

ORIGINATION AND PRE-PRESS
- Predict the level of shrink
- Digitize the subsequent distortion
- Provide a 3D preview
- Display the artwork distortion on the sleeve
- Pre-distort the artwork to compensate for production

Figure 1.15 Distortion is the major challenge

market players and is forecast to gain further momentum with future manufacturing investment.

Finally, Latin America and some parts of Africa are the most recent to adopt shrink sleeve labeling technology. The growth of the market in these regions is expected to be somewhat sluggish between 2017 and 2021.

With the market penetration of shrink sleeves in Asia Pacific being very high, the industry has its sights set on achieving the same level of penetration in Europe and North America as in Asia Pacific. To provide some context for the growth rate of shrink sleeves in North America, more specifically in the United States, Figure 1.11 (derived from industry sources) illustrates the size of the shrink sleeve market in the year 2000 and again in the year 2014. In this 14-year period, the shrink sleeve market grew by almost a factor of ten!

In looking at the historical and forecast growth of the global sleeve label market to 2020 (Figure 1.12), we observe that heat shrink sleeve labels (the blue bars) provide the dominant sleeve label technology growth. The stretch sleeve growth is shown in red, and the Roll-on shrink-on (R.O.S.O.™) and other technologies are the green bars at the top. With projected sales of over 10,500,000,000 square meters by 2020, heat shrink sleeve labels are the focus of the labeling industry for very good reason.

FILMS USED FOR SHRINK SLEEVE LABELING
Films are quite important to the success of shrink sleeve labeling. Different films shrink at different rates to suit various applications, container shapes, and different container materials.

Figure 1.13 details the types of shrink film

materials available in the market. We will discuss these films in greater detail in the chapters ahead, but the most important takeaway for now is to understand that film selection will depend on the application, the shape and size of the container we are decorating, the material of which the container is comprised, and other factors we will explore as we progress through this book.

SLEEVE PRODUCTION
In order to produce a shrink sleeve decorated container of the highest quality, we must understand the various production stages, from origination and pre-press through printing, slitting, tube forming, tube application and shrinking to achieve a tight fit on the container. These stages are summarized in Figure 1.14.

A key challenge associated with shrink sleeve decoration is the origination, or the design and image preparation. The origination must be created with an element of distortion that predicts the level of shrink such that when the printed film is shrunk onto the container, it will appear proportionate and not misshapen (Figure 1.15). The pre-press step of the shrink sleeve process was therefore complex, labor-intensive, and it required plenty of trial and error, with designs created two-dimensionally in anticipation of how they would shrink onto a three-dimensional object.

Fortunately, advances in pre-press software technology provide users with realistic results, demonstrating how the artwork will distort during simulated trials and what the finished product will look like, sparing converters the hours and hours of trial and error that once used to be a key component of the shrink sleeve process. We will discuss this in greater detail in Chapter 3.

Following our discussion of origination and pre-press in Chapter 3, we will discuss the printing and the ink selection processes in Chapter 4. Flexography is forecast to be the most preferred printing technology in the coming years, but rotogravure and offset lithography are still widely used. Digital printing has gained increased popularity as of late, both with toner and inkjet technologies. In a similar vein, ink chemistries vary and will behave differently based on the materials on which they are printed. The ink chemistry needs to 'stay' with the film as the film is shrinking in the shrink tunnel.

After printing, the web will be slit to the correct width, where it will be ready to be formed, or seamed, into a tube. Seaming is the process of welding the edge of the shrink materials together with a solvent to form the sleeve before it is cut or rewound onto a roll. Seam location will vary from one shrink sleeve to another, depending on variables like container shape and style, automatic or hand application and artwork design. Slitting and seaming has its own terminology, such as 'layflat', 'slit width', 'seam location', 'solvent width', 'U-fold', 'skips and voids', and 'solvent control'. All of these terms, together with the technology of folding and forming, rewinding and oscillation, monitoring and inspection, are discussed in greater detail in Chapter 5.

Application and heat shrinking are the final links of

the shrink sleeve chain. All of the chapters up to this point will look at how to produce a high value, high quality label. However, it is during the application part of the process where some of the key challenges lie, making this stage as important as the previous steps. We will also introduce some new terminology, like 'overlap', 'repeat length', 'cut length', 'clear area', 'curl back', and 'smiling and frowning', all of which require some depth of understanding before examining the various types of sleeve application machinery, sleeve manipulation solutions, and shrink tunnels and accessories. All of these aspects will be covered in Chapter 6.

NEW DEVELOPMENTS

Innovation continues to expand and develop the shrink sleeve label market. Some recent advances include color change and glow-in-the-dark sleeves, limited edition and personalized sleeves, hidden prizes and coupons, multipacks, biodegradable films and co-extruded films, lightweight containers (for cost-saving purposes), microwaveable sleeves, and advances in shrink tunnel technology. Additionally, the sleeve industry has introduced full-length container decoration on complexly-shaped containers, combined body label and tamper evidence, 360-degree decoration, a wide range of finishes such as matte, gloss and pearlized, reduced wall thickness of plastic containers, and UV barrier properties.

To summarize, the shrink sleeve industry remains as dynamic and vibrant as ever. With so much growth potential in virtually every part of the world, the opportunities that this label technology presents to converters – not to mention the marketplace in general – are virtually limitless. Brand owners are eager to employ this technology to share their brand's story with consumers at the point of sale, and consumers are drawn to the shapes, designs, colors and textures that shrink sleeve labels bring to life on the containers of products that consumers purchase. Shrink sleeve labeling will continue to inspire the market for decades to come.

Chapter 2

Shrink sleeve substrates and their usage

Chapter 2 provides an overview of shrink sleeve substrates, their uses, and how to choose them. This chapter will not have hard and fast rules, but will instead provide advice and guidance to practitioners as they embark on the shrink sleeve label production journey.

Heat shrink substrates are comprised of a number of polymers. The most common include OPS (Oriented Polystyrene), PVC (Polyvinyl Chloride), PETg (Polyester) and Polyolefins. Additionally, combination films, Hybrid or Multilayer films, and speciality films like White Opaque (Figure 2.1) also round out the substrate offerings. This list is not exhaustive, but it does give a sense for the breadth of options that exist. Each film has different properties and characteristics that require consideration when choosing the most appropriate solution.

In addition to substrates, we must also consider certain container characteristics and even

PVC	**White Opaque**
OPS	**PETg**
Polyolefin	**Hybrid/Multi-layer**

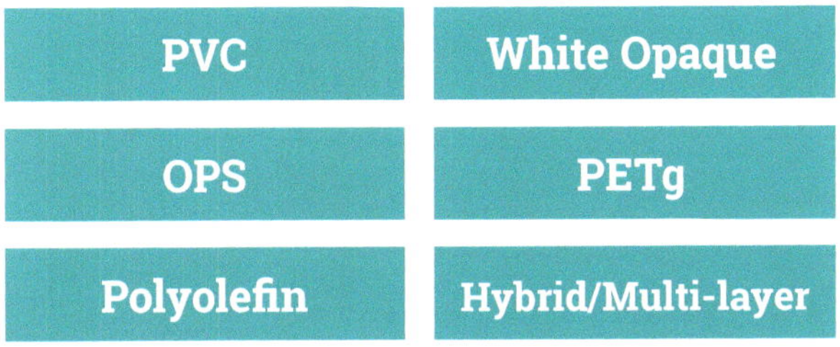

Figure 2.1 Some of the polymers used for shrink film labels. Source: Klöckner Pentaplast

Squeezability
Shrink Force
Print Quality
Max Shrink
Natural Shrinkage
Smiling and Frowning
Scuffing
Relaxation

Figure 2.2 There are many factors to be considered in film selection. Source: Klöckner Pentaplast

environmental settings (Figure 2.2). For instance, we must consider **squeezability**: is the container squeezable, and which substrate should be chosen to make it easy to squeeze? What about scuffing?

How does the material perform when applied to the final product and is transported? Is it going to be packaged as a half tray bundle wrap; or is it going to be surrounded by cardboard protection? How does the shrink force affect the choice of film? These are just a few examples of what must be considered when choosing which substrate to use.

There are other general considerations as well, like light sensitivity: if the product is sensitive to light, we must factor this variable into the type of substrate we select. Similarly, if the product experiences rough handling from the production facility to the store shelf, we must take this into consideration when selecting the appropriate substrate. In short, there is a range of factors to consider when choosing the appropriate substrate.

CONTAINER CHARACTERISTICS

The starting point for selecting a shrink film is the container shape and characteristics. The container shown in Figure 2.3 is a trigger bottle, an asymmetric shape with long sides. This type of container can be round, it can be square, or it can be long-sided. Each container type has different characteristics that will determine the shrink sleeve substrate most appropriate for the container.

Determining the appropriate substrate means starting with some fundamental questions, which we will divide into four categories (Figure 2.3). Maximum shrinkage percentage is critical. The film needs to shrink enough so it does not show 'flowering', an effect that occurs when the film does not shrink

enough, causing the top or the bottom to 'flower.' The substrate must shrink enough to provide complete coverage of the container being decorated. For example, if the sleeve is not required to go to the top of a narrow neck, it will require a lower shrinkage than one that does.

Is a high or low shrink force required? Shrink force is another significant consideration. Shrink force is the force exerted on the container as the film shrinks onto it. Shrink force factor considerations, such as shrink tunnel type and shrink tunnel effects, will be discussed in greater detail later in this chapter.

Now we will address these four fundamental questions one by one.

Maximum shrinkage. There are several factors to consider regarding maximum shrinkage. One factor is the container shape and the maximum shrinkage required to cover it. Other considerations include the container contours, the container wall thickness, whether the container is full or empty and the tunnel type. Does the container have ribs? Is it a trigger bottle? Is it asymmetric?

The formula for the maximum shrink required is shown in **Figure 2.4.** Simply take the narrowest circumference (or diameter for a round container), divide by the widest circumference (or diameter), subtract from 1, and change to a percentage. Then add 1%-2% for layflat oversize and safety. This gives the maximum shrink percent required of the substrate.

Indeed, what is layflat oversize and why do we need it? The layflat of a shrink sleeve is when a shrink sleeve reel has been made into a tube and then is

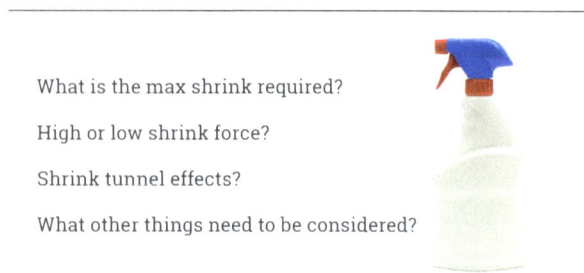

What is the max shrink required?

High or low shrink force?

Shrink tunnel effects?

What other things need to be considered?

Figure 2.3 Container shape and the fundamental questions to ask. Source: Klöckner Pentaplast

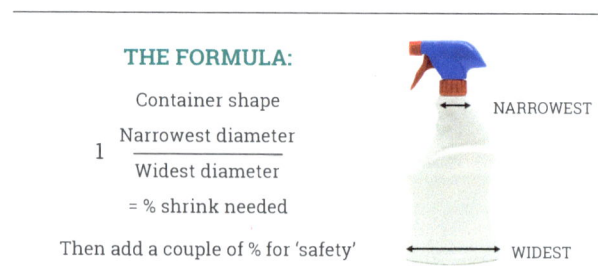

THE FORMULA:

Container shape

$$1 - \frac{\text{Narrowest diameter}}{\text{Widest diameter}}$$

= % shrink needed

Then add a couple of % for 'safety'

NARROWEST

WIDEST

Figure 2.4 The formula for determining maximum shrinkage. Source: Klöckner Pentaplast flattened. Measuring from one side to the other of the flattened tube is called the layflat. (More information on this can be found in Chapter 5 and using Figure 5.18). If the layflat is exactly the same as the circumference of the container, the sleeve will not fit over the container smoothly and will jam. The machinery application manufacturer will advise what layflat oversize is needed; for instance, it could be 5mm, or 10mm or so. It depends on the container shape and how the shoulder is formed. For instance, if the container has a sharp shoulder, then the layflat will need to be a little bit larger. We must also consider the speed at which the shrink sleeve label is applied; when running faster, more oversize will be needed. For example, if the speed is 400 bottles per minute, the applicator supplier will advise that a specific layflat oversize is required, and this specific layflat oversize enters our calculus.

The formula then provides the total percent shrink needed, which may be, for instance, 60% shrink. Always purchase a film with slightly more shrinkage than the total percent shrink needed to make sure it completely shrinks over the container. In the example provided, you would purchase a film with a shrinkage of, say, 62%.

Shrink force. How does shrink force affect the choice of substrate? How about ribs or contours? Let's assume the container has ribs. A weak shrink force won't pull the film into the ribs; it will just 'bridge over' them. When the consumer then handles that container, it will feel cheap and ill-fitted. They will be able to feel the loose film over the ribs. To get a good customer feel, the film should pull into the ribs and be tight on them – so a higher shrink force is required.

However, if the container is empty, particularly a weak container (for instance, a HDPE container that is being sleeved empty), then a high shrink force will distort the bottle in the shrink tunnel. When the film shrinks onto the container, the film will be strong enough to twist and distort the bottle. This could mean that the bottle cannot be filled, and cannot be capped. In short, container wall thickness, tunnel type, whether the container is being sleeved empty or filled, all matter in the choice of substrate.

Shrink tunnel effects. Relaxation is another important consideration (Figure 2.5). As an empty

'Relaxation'

Smiling/Frowning

Wrinkles

SMILING

FROWNING

Figure 2.5 Some of the key shrink tunnel effects
Source: Klöckner Pentaplast

HDPE container travels through a shrink tunnel, it will be in the shrink tunnel for maybe five or 10 seconds, depending on the length of the tunnel, speed of the line, etc. The container will expand as it goes through the tunnel due to the heat. When the label shrinks onto the expanded container inside the tunnel, the label will fit the container snugly while inside the tunnel. However, when it comes out of the tunnel and returns back to ambient temperature, the container will shrink back.

What then happens to the sleeve? It does not shrink back, and the sleeve does not fit tightly anymore – it becomes loose and can spin on the container; the film does not 'relax' back with the container. This loose fit does not provide a good customer feel. What is needed is a film that shrinks back with the container as it cools. As the container reduces in size, the film needs to shrink with it and remain tight. This is referred to as 'relaxation' - the film needs to 'relax' with the container.

Now, let's discuss **smiling/frowning**. Consider a long-sided oval container, a square container, or a trigger container. If the wrong material is chosen, the shrink sleeve will 'pull up' from the bottom on the long side of the container. This effect is called 'frowning'. A similar effect happens at the top of the container – the label will pull down and cause a 'smile', which appears unsightly on the retail shelf. The converter does not want a label with this effect, and the brand owners certainly do not like it. There are certain materials that have a tendency not to frown or smile, but there are others that do; choosing the right substrate for these shaped containers is important. If it is a round container, it is not as important, but on a container with a long, straight side, it can be critically important.

There are also a number of other things to consider and take into account when choosing the right substrate (see Figure 2.6).

Does the brand owner accept PVC? PVC was the first material developed for shrink sleeves. It is considered a great shrink sleeve substrate in a number of ways. However, some brand owners consider PVC a problem due to environmental considerations. Therefore, one of the first considerations will be: 'What is allowed by the brand owner? Will the brand owner accept PVC?' This is a fundamental question that affects substrate choice.

Light blocking: Certain drinks require protection from the effects of light, which therefore becomes another consideration when selecting the appropriate substrate. Some examples include dairy beverages, isotonic drinks, and aseptically packaged drinks. In addition, product constituents such as vitamins, particularly riboflavin, also require protection from the effects of light. Finally, taste and flavor attributes require protection from the effects of light. In summary, visible light and/or UV light can destroy or change the nature of the product being decorated, and we must enter these factors into the substrate equation.

Ultraviolet detection: Some application machines use ultraviolet detection to determine whether the sleeve was applied correctly to the container. For this reason, the sleeves need an optical brightener present in the film. The application machine 'sees' if the sleeve is positioned correctly on the container, and if it is not, it will divert the container off the conveyer. It is possible for the printer to print the optical brightener (OB) onto the substrate, but if the print levels of OB fade, there is potential for large rejections of sleeves – so having the OB in the substrate is more reliable.

Transportation: In the United States, for instance, many beverages are transported using half trays with bundle wrap. The containers are packed closely together and often rub against one another during transport. Because of this, the shrink sleeves on the containers can be scuffed or damaged, and the result is an unsightly product. So a consideration in the choice of shrink film substrate is how resistant to scuffing must the film substrate be?

Recyclability: The shrink sleeve itself generally cannot be recycled into clear recycled polymer as they are printed; they can only be recycled into secondary uses. However, the container generally can be recycled. Hence, a consideration in the choice of substrate is how the sleeve will be separated from the container in the recycle system? The sleeve needs to be removed from the container so that the container (and the sleeve separately) can be recycled. This has become an increasingly important consideration recently as ever more containers are sleeved.

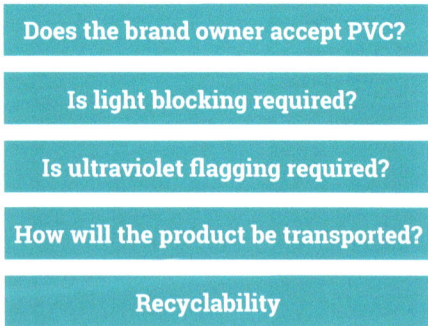

LABEL FILM TYPES AND CHARACTERISTICS

Having talked about containers and their characteristics, we will now start looking at the range of films that can be used and their characteristics. What substrates are best used for which applications? We shall look at the key characteristics of each shrink film and their main technical details.

PVC (Polyvinyl Chloride)

PVC is the oldest of the shrink films and a terrific all-around film. This film processes well, it prints well, it shrinks well and it has a great shrink curve. It can be printed with rotogravure, letterpress, offset and flexo, using UV, solvent, or water-based inks. It is a great

| Does the brand owner accept PVC? |
| Is light blocking required? |
| Is ultraviolet flagging required? |
| How will the product be transported? |
| Recyclability |

Figure 2.6 Some of the fundamental things that need to be considered. Source: Klöckner Pentaplast

material. It can be supplied in UV protection grades, as well as with optical brightener grades. It has been around for so long that it is possible to get PVC film in a host of variations. It is available with low shrink, medium shrink and high shrink, but, as mentioned before, the primary concern with using this film is whether the brand owner will accept it.

As can be seen in Figure 2.7, PVC shrink is midrange at 55-65%; not particularly high, though not low.

PVC has a moderate shrink force that will crush

PVC (Polyvinyl Chloride)

CHARACTERISTICS

'Tried and true': PVC is the traditional mainstay of film labels

Processes and handles well

Accepts all types of printing

Can meet UV protection & optical brightener needs

May not conform to brand's environmental standards

TECHNICAL DETAILS

Shrinkage:	**55%-65%**
Shrink force:	**Moderate**
Relaxation:	**Poor**
Smile/frown resistance:	**Poor**
Scuff resistance:	**Fair**
Squeezability:	**Poor**
Natural shrinkage:	**Good**
Printing:	**Good**

Figure 2.7 Characteristics and technical details for PVC film. Source: Klöckner Pentaplast

---

empty containers. It will also not 'relax' on empty HDPE containers, resulting in a loose sleeve. The scuff resistance is middling – better than OPS, but not as good as PETg. Finally, squeezability is poor; on a squeezable container, the sleeve will crinkle and will have a poor customer 'feel'.

Another term worthy of mention is **natural shrinkage**, which is a term that pertains to converters. When a roll of film is purchased it will be stored in the converter's warehouse, where it will gradually shrink (creep), depending on the warehouse temperature. Usually, the shrink is imperceptible. PET hardly shrinks at all; OPS will shrink significantly more. Films that have greater levels of natural shrinkage likely will have shrunk by the time they are put on the press. After printing, the film will move into a warehouse, where it may continue to shrink, unless the warehouse is kept very cold. If the converter is in a hot place, OPS can be very difficult to manage. Natural shrinkage for PVC is good, i.e., it has very little natural shrinkage when kept at reasonable warehouse temperatures. Natural shrinkage concerns material that is printed and handled; in other words, once it has been shrunk onto the container, natural shrinkage no longer becomes a concern.

PETg (Polyethylene Terephthalate, Glycol-Modified)

PETg is typically referred to as just PET. However, there is no such thing as a shrink sleeve made from pure PET. Every PET shrink sleeve is made from a PET that has been modified in some way. The general terminology we use for this modified film is PETg. It is a clear film and it has the highest maximum shrinkage of all shrink film substrates. If a converter desires shrink rates of 75% or greater, this is the only substrate option. It will shrink all the way down to the neck of a narrow neck beer bottle or two liter beverage bottle. It is commonly used when PVC is not an option allowed by the brand owner. It handles well and has an excellent printing surface – printing really stands out and has a vibrant look. When comparing PET and PVC side by side, the PVC prints well, but PET will stand out a bit more. PET substrates can also come with UV protection to protect the label or, at a higher level of protection, protect the product contents. It is also possible to get

PETg (Polyethylene Terephthalate Glycol-Modified)	
CHARACTERISTICS	
Clear film	
Has highest max shrinkage of all films	
Commonly used when PVC isn't an option	
Processes and handles well	
Excellent printing surface	
Can meet UV protection & optical brightness needs	
TECHNICAL DETAILS	
Shrinkage:	75%+
Shrink force:	Moderate to high
Relaxation:	Poor
Smile/frown resistance:	Depends on grade
Scuff resistance:	Excellent
Squeezability:	Poor
Natural shrinkage:	Good
Printing:	Excellent

Figure 2.8 Characteristics and technical details for PET shrink film. Source: Klöckner Pentaplast

White Opaque (non-clear PETg)	
CHARACTERISTICS	
Several versions on the market	
Glossy and matte options	
Needs surface printing & over-print varnish	
Black reverse (printed or bi-color) needed for total light blocking	
Some matte versions are floatable	
TECHNICAL DETAILS	
Shrinkage:	75%+
Shrink force:	Moderate to high
Relaxation:	Poor
Smile/frown resistance	Depends on grade
Scuff resistance:	Excellent
Squeezability:	Poor
Natural shrinkage:	Good
Printing:	Generally good, but matte versions sometimes have challenges in rotogravure

Figure 2.9 Characteristics and technical details for white opaque film. Source: Klöckner Pentaplast

optical brightener versions of PET for sleeve detection. It works well in steam tunnels, which we will discuss later. And if used with care, it can also operate in hot air tunnels.

Maximum shrinkage for PET is 75% or greater. The shrink force, however, is high. It will crush empty containers and distort them. It also will not relax, so if there is an empty HDPE container that has expanded during the shrink tunnel process, then a PET sleeve will not shrink back with the container, and will instead be loose. However, it is a tough, durable material and has the best scuff resistance. When two containers are placed together and rubbed, they will still appear unblemished, even after being shipped over long distances. For glass or metal containers, use PET. Glass next to glass will scuff – therefore a PET sleeve is required. Natural shrinkage for PET is good, and as mentioned before, PET is excellent for printing.

One thing worthy of mention about PET, however, is concerning smiling and frowning. There are several grades of film available, and some of them are specifically designed not to frown or to smile. The converter therefore has to be a bit careful about which PET to use with long-sided, non-round containers.

White Opaque (non-clear PETg)

There are several versions of white opaque film in the market that are PET films containing various mineral fillers. It is possible to purchase these in both glossy and matte versions. If light protection is needed, white opaque will provide about 70% to 80% light protection. To get 99% - 100% light protection, however, the converter will need to print black on the reverse of the film. Examples of this include products such as Nesquik or Fairlife dairy beverages in the U.S. that, when the label is removed, show a 100% black print on the inside of the sleeve. It is worth mentioning that the printer should not print the black over the seam, though. The black ink is printed to protect the contents from light. If the white opaque has been chosen only for aesthetics, though, there is no need to print the reverse black.

White opaque films offer shrinkage percentages that are very similar to those of clear PET – roughly 75% shrinkage.

New to the market are the 'bi-color' films – which are white on one side, and black or grey on the reverse side. These films provide complete light-blocking without requiring the printer to print the 100% black ink reverse print. Examples are the 'Eklipse' product from Klöckner Pentaplast.

Reference also needs to be made about the printing of white opaque films. White glossy film prints much the same as PET – very well. White opaque generally prints well unless it is printed rotogravure. The white opaque surface is not flat, which causes a matte effect. Shrink sleeves are commonly printed rotogravure: Asia, for example, is 100% rotogravure; Europe is maybe 50%; in North America, in the shrink business, it is around 30-45%. When the matte surface connects to a rotogravure cylinder in the printing process, the ink needs to come out of the tiny cells in the cylinder and transfer to the substrate. Any film surface variation will cause a problem with a fine 10% rotogravure halftone – the ink will not come out of the cells cleanly. The capillary action needed to pull the ink out of the gravure cell does not happen,

OPS (Oriented Polystyrene)	
CHARACTERISTICS	
Clear film with niche uses	
Less 'vibrant' printing than PET	
Tricky to process; very delicate handling required	
Needs care when printing with solvents: rotogravure printing requires less acetate as it attacks OPS	
Good in hot air tunnels	
TECHNICAL DETAILS	
Shrinkage:	55%-65%
Shrink force:	Very low
Relaxation:	Excellent
Smile/frown resistance:	Excellent
Scuff resistance:	Poor
Squeezability:	Excellent
Natural shrinkage:	Poor
Printing:	Needs care

Figure 2.10 Characteristics and technical details for Oriented Polystyrene. Source: Klöckner Pentaplast

causing a missing dot. So with half tones and highlights of 10%, 15% or 20% dots, there will be printed image problems. Electrostatic Assist (ESA) helps but does not solve the problem entirely. Moreover, many printers are wary of ESA with solvent gravure on safety grounds. Offset or flexo printing typically does not encounter issues of this sort when printing on white opaque film.

OPS (Oriented Polystyrene)

The word 'oriented' in OPS simply means it has been stretched. OPS is a great material. Nevertheless, the converter needs to be familiar with how to process it. It is tricky to print, particularly using rotogravure. It will

PO (Polyolefin)

CHARACTERISTICS

Clear, floatable film with moderate clarity

Prints reasonably well; needs corona treatment

Generally chosen for APR (US) or EPBP (EU) recyclability shrink sleeve standards for PET containers

Sometimes chosen for other properties, such as anti-crinkling

TECHNICAL DETAILS

Shrinkage:	**50%-60%**
Shrink force:	**Low**
Relaxation:	**Good**
Smile/frown resistance:	**Excellent**
Scuff resistance:	**Poor**
Squeezability:	**Excellent**
Natural shrinkage:	**Fair (but needs care)**
Printing:	**Good**

Figure 2.11 Characteristics and technical details of Polyolefin shrink film. Source: Klöckner Pentaplast

tend to shrink both in the warehouse and after it has been printed on and is sent to the customer. It is very susceptible to shrinking at even slightly elevated temperatures.

Care needs to be taken with OPS film – it is no match for oil and chemicals. Solvent rotogravure generally uses a mixture of acetates and alcohols, with a high percentage acetate and a low percentage alcohol. Unfortunately, if a drop of ethyl acetate is put on an OPS film, it will melt straight through the film, burning a hole through it. This shows how weak OPS film is against acetates, and yet the solvent

rotogravure converter would like to print with them. To print OPS, it is necessary to reverse the percentage and have an alcohol-rich ink system – requiring the ink in the press to be removed and replaced with a special OPS ink. When the converter goes back to a PVC or PET film, then the OPS ink will need to be removed and the press replenished of other inks. There are combination inks that can print both PET and OPS, but using these inks means carefully choosing your substrates.

OPS provides a medium to reasonably high shrinkage, as noted in Figure 2.10. It is possible to get OPS that goes higher than 65%, but it is a special OPS.

Relaxation is excellent for an empty HDPE container going through a shrink tunnel decorated with OPS substrate. OPS is very good at relaxing back, so if the bottle has expanded and the OPS shrinks onto it, then as the bottle cools and contracts, the OPS will just shrink back (relax) with it – it remains tight. However, rub two OPS sleeved containers together, and within three or four rubs it will show scuffing. Within 10 rubs, there will be a hole in the film. It is a very weak film, so if the decorated containers are going to be transported over a long distance, it is not advisable to use OPS.

In terms of squeezability, there is no crinkle. Squeeze OPS and it returns to its shape, it takes the form back to that of the expanded bottle. Brand owners love it for this reason, but it requires care when it is handled and printed.

Polyolefin (PO)

Polyolefin shrink sleeve film is generally a combination of polyethylene (PE) and polypropylene (PP).

The most common labeling technology used to label PET beverage bottles is the polypropylene (PP) wrap-around label. The recycling industry has developed a straightforward technology to separate the PP wrap-around label from the PET container using float tanks, where, after grinding, the PET bottle flakes sink and the PP label floats. In doing so, they separate the valuable, clean PET bottle flakes from the printed PP label, which allows the PET bottle flakes to be recycled. Unfortunately, the specific gravity of PETg, PVC and OPS shrink labels is greater than 1.0, meaning that they will sink in a float tank.

Hybrid/Layered (Multi-Polymer)	
CHARACTERISTICS	
Clear film combining advantages of PET & OPS	
Excellent printing surface	
Good in hot air tunnels	
TECHNICAL DETAILS	
Shrinkage:	~65%-70%
Shrink force:	Low
Relaxation:	Good on HDPE containers
Smile/frown resistance:	Very good
Scuff resistance:	Excellent
Squeezability:	Good
Natural shrinkage:	Fair
Printing:	Excellent

Figure 2.12 Characteristics and technical details of Hybrid/Layer shrink sleeve films. Source: Klöckner Pentaplast

The advantage of polyolefin shrink sleeves is that they float, however. The polyolefin shrink sleeve material can therefore be separated from the PET bottle flakes, allowing the PET bottle flakes to be recycled.

PO films are also good on crinkle and relaxation. Their performance rivals that of OPS, and because of this, brand owners' interest level in PO is quite high.

Hybrid/Layered (Multi-Polymer)

OPS film has many drawbacks, particularly in processability and scuff resistance, as detailed in Figure 2.10. However, there is a material available that overcomes these drawbacks. Such film consists of an OPS core with a PET skin on each side. This provides the gentle shrink characteristics of OPS, but adds the printing, robustness and scuff resistance benefits of PET, providing the best of both worlds. This is the Hybrid/Layered multi-polymer film.

Shrinkage of these films is medium to high at 65-70%. It does relax well, similar to OPS. Scuff resistance is excellent and it is straightforward to print. It is the perfect material in many ways. Natural shrinkage is not as bad as OPS but not quite as good as PET; printing with a PET surface is excellent. Hybrid/Layered films comprise the best elements of PET and OPS and are therefore an excellent option.

SUMMARIZING THE SHRINK SLEEVE FILM CHARACTERISTICS

Having reviewed the main types of shrink sleeve films, their characteristics and technical details, we may now put all of this together in one simplified summary table (Figure 2.13). It is a great visual in terms of showing the pluses and minuses for each of the materials and their technical considerations. It should be mentioned, however, that these are guidelines – not hard-and-fast rules – and also are not the only considerations.

UNDERSTANDING THE DECISION-MAKING JOURNEY

To this point in Chapter 2, we have discussed the effect of container shapes, shrink process requirements and the characteristics and technical details of the various shrink films. The aim now is to better understand the film selection decision-making process, provide a guide to the decision process itself and then to look at a number of example decision-making journeys.

To begin this journey, it will help to understand further the nature of the choices being made using Figure 2.14. We need to start narrowing down the shrink film options. First, we will start with the top row of the chart: the shrink tunnel. Is it a hot air or radiant heat tunnel? Is it a steam tunnel?

Steam tunnels are very common, and as indicated by the name, they dispense steam through nozzles on each side of the tunnel as the container travels through. The steam creates a two-phase system,

SUMMARY OF FILM CHARACTERISTICS

	SHRINKAGE	SHRINK FORCE	RELAXATION	SMILING	SCUFFING	SQUEEZING	NATURAL SHRINKAGE	PRINTING
PVC	55%-65%	Moderate	Poor	Poor	Fair	Poor	Good	Good
PETg (inc. White Opaque)	75%+	Moderate-High	Poor	Depends on Grade	Excellent	Poor	Good	Excellent
OPS	55%-65%	Very Low	Excellent	Excellent	Poor	Excellent	Poor	Needs Care
HYBRID/ LAYERED	65%-70%	Low	Good	Very Good	Excellent	Good	Fair	Excellent
PO	50%-60%	Low	Good	Fair	Poor	Good	Needs Care	Good

Figure 2.13 A summary of the characteristics and technical details of shrink sleeve films. Source: Klöckner Pentaplast

condensing on the sleeve while the energy from the latent heat of vaporization goes into the sleeve. The steam envelopes the containers and condenses evenly on the sleeves, creating a very even shrink. As it condenses on the film, it shrinks gradually, but with a controllable shrink. Steam shrink tunnels are very effective, but they are not used in powder plants, in applications where steam cannot be used, or where steam is not available. Brand owners also tend to be wary of using steam tunnels when sleeving empty containers.

Hot air tunnels. Hot air tunnels direct hot air jets at the shrink sleeve to shrink it uniformly but, as can be imagined, a jet of hot air is not distributed evenly, so will tend to shrink the sleeve unevenly. To counteract this, the shrink films are designed with a very smooth, gentle shrink curve characteristic where the material will perform even though it is being impinged with different temperatures and different velocities. In this way, a good shrink result can be created with a hot air tunnel. Radiant tunnels have a more uneven temperature and shrink distribution than do hot air tunnels but are generally used for tamper evident applications only.

The next row down looks at the **maximum shrink percentage**, which was discussed earlier in the chapter. This creates another way of narrowing down the field of potential shrink films for a particular application.

The third row down has to do with container type. Is it a glass container? A metal container? An HDPE container that is cold and filled – a milk chug, for instance? Or is it an HDPE container that is empty or hot, i.e., it is going to shrink back and potentially create a loose sleeve? Is the container a PET container, cold and filled, or an empty PET container? This row details the container that is to be sleeved and its condition.

The next row reviews some additional requirements. What about scuffing – is this a concern? Is PVC an option (i.e., allowed by the brand owner) or not? Is floatability required for separation of the sleeve from the container? What about the container shape? Does it have ribs or a tightly contoured container where a high shrink force is needed to pull it in?

The bottom row of the diagram lists out the different substrate options.

DECISION JOURNEY: WE HAVE A WINNER...

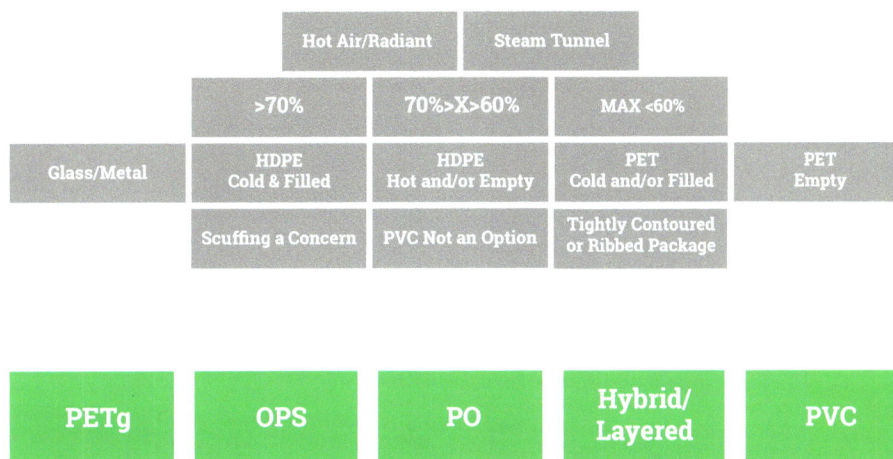

	Hot Air/Radiant	Steam Tunnel		
	>70%	70%>X>60%	MAX <60%	
Glass/Metal	HDPE Cold & Filled	HDPE Hot and/or Empty	PET Cold and/or Filled	PET Empty
	Scuffing a Concern	PVC Not an Option	Tightly Contoured or Ribbed Package	

PETg	OPS	PO	Hybrid/ Layered	PVC

Figure 2.14 The decision making journey. Source: Klöckner Pentaplast

Now that the choices have been defined, it becomes possible to work through the following decision-tree examples that will illustrate the process.

Example 1

We begin with Figure 2.15 and select a hot air tunnel in the top row. The selection of a hot air tunnel results in PET film being removed from the options (note that films taken out are shown in red). Why? It is possible with particular PET films and particular container shapes to make PET work well in a hot air tunnel. Nevertheless, PET struggles to make a good shrink result in a hot air tunnel – the shrink curve is too steep, and it shrinks too quickly. So PET with a high shrinkage is discounted as an option.

However, if the container requires a very high shrink film, then it is possible that PET may have to be used – and getting a good shrink result will be difficult. Therefore, the material needs to be chosen very carefully. What if the max shrink required is less than 60%? A very low shrink film will be needed and all other options remain viable.

At the next level down, assume in this scenario that the container is a HDPE cold filled milk container that needs a medium shrink (i.e., between 60% and

70%). Everything aside from PET (which we lost above) stays in and nothing else drops out. What else can be considered? If we assume that the brand owner doesn't accept PVC, we highlight that box and PVC drops out. We can also assume that these containers are going to be transported cross-country in half-trays, so scuffing is a consideration. This requirement eliminates PO film and OPS film from the list of viable options. This results in the ideal material being a Hybrid/Layered film. Please note, this decision tree says nothing about price; this decision tree only factors technical considerations.

Other materials can potentially work in this particular scenario; but the most suitable substrate for this particular situation is the Hybrid film (Figure 2.15).

Example 2

The tunnel is hot air again and the container requires between 60% and 70% shrinkage. The container in this example is a PET container that is cold and filled. PETg will therefore be removed as an option due to the hot air tunnel.

The gentlest shrink curve comes from OPS film, but it is highly susceptible to scuffing. If scuffing is not an issue, OPS is a great option. However, as in the

DECISION JOURNEY: WE HAVE A WINNER...

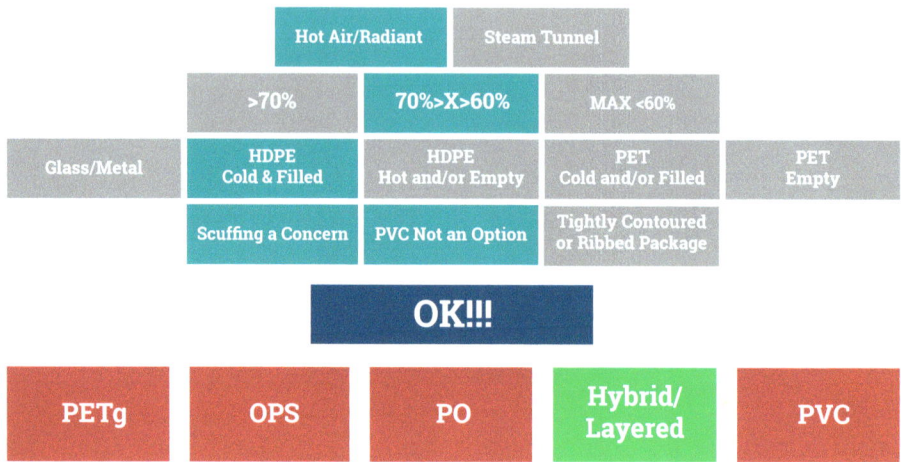

Figure 2.15 Shrink film choice for a hot air tunnel with 60% -70% shrinkage, a cold-filled HDPE container, and where scuffing is a concern and PVC is not preferred by the brand owner. Source: Klöckner Pentaplast

DECISION JOURNEY: WE HAVE A WINNER...

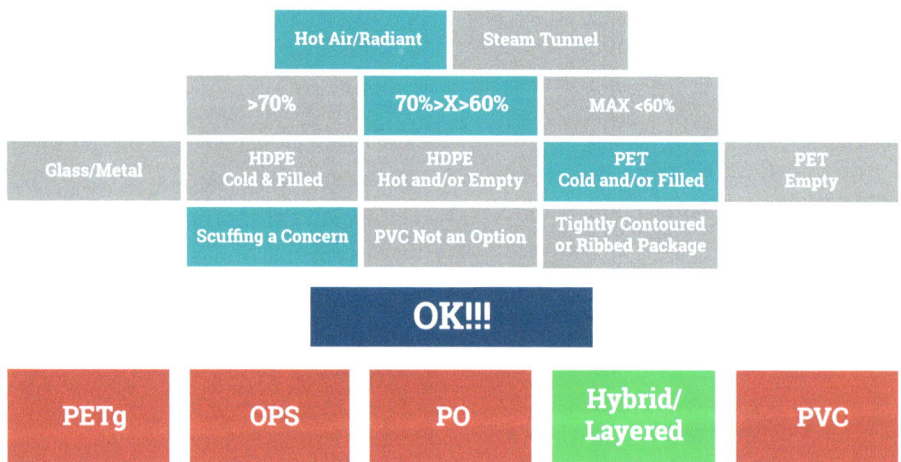

Figure 2.16 Shrink film choice for a hot air tunnel with less than 60% shrinkage, a cold and filled PET container and where scuffing is a concern. Source: Klöckner Pentaplast

DECISION JOURNEY: WE HAVE A WINNER...

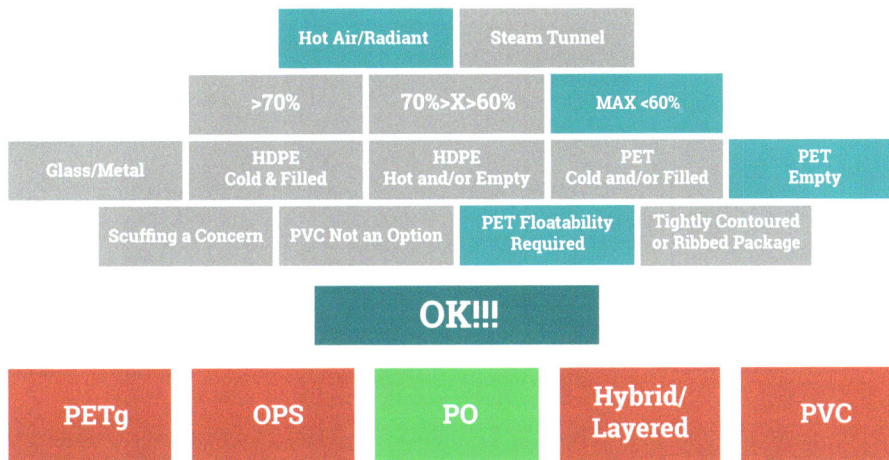

Figure 2.17 Shrink film choice for a hot air tunnel with less than 60% shrinkage, an empty PET container and where floatability for recycling is required. Source: Klöckner Pentaplast

diagram, if scuffing is a concern, then PVC, OPS and PO all will drop out, and again, the ideal solution will end up as a Hybrid film (Figure 2.16).

Example 3

This time we are using a hot air shrink tunnel (therefore, no PET film), the container requires less than 60% shrinkage, and we are shrink labeling an empty PET container. PVC can be discounted because its shrink force is too high and the PVC sleeve will distort the container as it shrinks. We need a sleeve with very low shrink force, which leaves PO, OPS and Hybrid substrates as options. However, in this case the brand owner has requested a floatable shrink sleeve so that the sleeve can be easily separated from the PET bottle in the recycling system using a float tank. In this case, the only substrate that fits this purpose is the polyolefin (PO) shrink sleeve substrate (Figure 2.17).

Other considerations

Once all the factors have been taken into account and the heat shrink film has been carefully chosen, the result will be a beautifully shrink sleeved container. Having said that, the shrink sleeve tree that has been

used in this chapter is still rather basic, and additional questions, queries or issues may arise.

One such question might be that of whether corona treatment (see also Chapter 4) is required when printing the substrate. The only sleeve material on which corona treatment is always needed is polyolefin film. Depending on what the printing technology is and depending on the inks, the other substrates do not need corona, however it could be helpful, particularly when printing UV inks on PVC and PET. If, corona treatment is used, it should not be overdone, particularly on PETg. Over-corona can produce some dramatic effects – creating blockages or even blowholes in the substrate.

For polyolefin film, the film manufacturer will normally corona treat with a high corona level. When the converter uses corona during printing, only a little 'bump' is required to increase the dyne level of the material so that the ink wets out better.

PETg and PVC accept ink well without surface treatment – when printed UV flexo, solvent- or water-based flexo, solvent rotogravure, UV or EB offset.

Another question that sometimes arises is

concerning PLA (Polylactic Acid) shrink films. Polylactic Acid is a plant-based material, generally made from corn. Only one company produces the film, Plastic Suppliers, and the use of PLA is always brand owner driven. If the brand owner needs a PLA sleeve, they should contact Plastic Suppliers. There are pluses and minuses with PLA; contact this company for further information and to determine if this film is an appropriate solution.

What about the shrink film gauge? Shrink films are available in a variety of different thicknesses, but most applications use between a 40-micron and 50-micron film. For some applications, 60-micron or 70-micron is needed, but these are rare.

Much of the thickness determination comes from the application machine. If the brand owner wants to run their plants as fast as possible with a wide process window, then 50-micron is generally most appropriate. If the brand owner wants to save some money, go to a 40- or 45-micron: although this might be a bit more tricky to shrink – the process window of the shrink process will narrow, and the tunnel may be more difficult to set up for a reliable good shrink result.

How does film thickness affect the shrink force? The shrink force varies linearly with caliper; a thinner film has a lower shrink force. However, the shrink force difference between OPS and PET is much larger than the variation from thickness, so moving from 50-micron to 40-micron with a PET film will still be nowhere near as low as a 50-micron OPS. The OPS is a fundamentally softer, gentler material.

It is worthwhile to mention some guidelines regarding the price differential between different shrink films. Firstly, the density of the substrate needs to be considered – the printer and brand owner are more interested in the cost per square meter, than how heavy it is! The density of OPS is about 1.05 g/cm3. The density of PET is about 1.32 g/cm3. The density of PVC is about 1.35 g/cm3. PVC and PET are about the same density, but OPS is much less dense. OPS is more expensive per kg but as the yield is higher, you get more square meters for each kg of film. To compare film costs, it is necessary to calculate back to the square meter by using yield figures provided by the film manufacturer. Hybrid has a density that is in between that of PETg and OPS at about 1.1/1.15. Polyolefin is below 1 (it floats), coming in at around 0.95-0.97 g/cm3.

When comparing film prices, always consider the density/yield of the film. Be particularly careful with white opaque films, which can either be very heavy with low yield due to mineral fillers, or so heavily voided that they have very light densities – some even less than water – resulting in very high yields.

Suppliers

There are many companies globally that supply heat shrink substrates. United States-based suppliers are listed in Figure 2.18.

	KLÖCKNER PENTAPLAST	BONSET AMERICA CORPORATION	GUNZE PLASTICS & ENGINEERING CORP	SKC FILMS	PLASTIC SUPPLIERS, INC
SUPPLIER	KLOCKNER PENTAPLAST	BONSET AMERICA CORPORATION	GUNZE PLASTICS & ENGINEERING CORP	SKC FILMS	PLASTIC SUPPLIERS, INC
ADDRESS	3585 Kloeckner Road, Gordonsville VA 22942	6107 Corporate Park Dr. Browns Summit NC 27214	1400 S Hamilton Circle Olathe KS 66061	1000 SKC Drive Covington GA 30014	2887 Johnstown Rd. Columbus OH 43219
CONTACT DETAILS	Andrew Lewandowski a.lewandowski@kpfilms.com +1-540-832-1591 Lilia Pedroso l.pedroso@kpfilms.com +55-11-4613-9979	John Uhlman juhlman@bonset.com (336) 375-0234	John Kramer john.kramer@gunzepa.com Chris Ross chris.ross@gunzepa.com (913) 829-5577	Sung Jin Kim sjkim@skci.com (866) 752-3456	(614) 471-9100
PVC	■	■			
PETg	■	■		■	■
OPS		■	■		■
PO	■				
HYBRID	■		■		
WHITE OPAQUE	■	■		■	■
PLA					■

Figure 2.18 United States-based suppliers of heat shrink substrates. Source: Klöckner Pentaplast

Chapter 3

Design and origination for sleeve labels

Design, pre-press and container selection are key elements to consider in sleeve technology, especially when looking at the topic of 'distortion'. Distortion is a difficult process to understand as it is necessary to anticipate how the sleeve graphics will become altered when they go through a heat tunnel. This in turn means that the graphics will have to be counter-distorted in advance. There are also design considerations that need discussing, as well as things like dielines, artwork, barcodes, proofing and plates. However, before moving into the technicalities of origination and pre-press, it is necessary to step back and examine what the shrink sleeve label offers, to both the brand owner and the consumer.

Shrink sleeve technology enables a product to capture an audience: grabbing the attention of the consumer as he or she walks through the shop aisles and sees a beautiful, 360-degree decorated product. The goal is to have the consumer reach out for the shrink sleeve-decorated product and purchase it.

The brand owner is most concerned with creating and executing a shrink sleeve design that engages the customer at the point of sale. In order to execute a truly successful design, however, the brand owner must work closely with the structural designer, the graphic designer and the material supplier to select a container shape and design that meets the needs of the container filling line and can be applied and processed through a high speed shrink tunnel.

In spite of these considerations, the ability to utilize 360-degrees of the container's surface presents amazing marketing opportunities for the brand owner. Figure 3.1 provides a great illustration of the opportunity that this labeling technology presents.

Figure 3.1 Comparison between a label (left) and a sleeve (right) used to decorate Barcardi Breezer. Source: Esko

Bacardi, the well-known spirits company, changed their Breezer design from a label (on the left) to a shrink sleeve (on the right) and realized increased sales revenues by doing so. In this instance, the brand owner had utilized roughly 25 percent of the surface area of the bottle to convey the brand message. With the shrink sleeve, the surface area usage increased to almost 100 percent.

Figure 3.2 The Coca-Cola shape translates and communicates the brand. Source: Esko

Look and compare the two, and then make up your own mind. Which package is more compelling?

CONTAINER SHAPE AND SIZE

The size, the color and the shape of a product will determine the type of container to be used. Bacardi used an outstanding design and had the courage to select a shrink sleeve label to maximize product impact at the point of sale. However, the selection of the container itself, the primary package, should be a shape that the brand owner selects that translates and communicates the brand. The shape of the primary container becomes inseparable from the brand style and drives brand recognition, brand loyalty and ultimately product sales. A perfect

example of that is Coca-Cola (Figure 3.2).

In addition to aesthetics, shape can also incorporate function. It can be ergonomic, easy to grab, to hold on to. For these reasons, it is important to take a holistic and inclusive approach to the design process. Certainly, the design process needs to be discussed early on. What kind of shape is going to be used? This is typically a selection made by the brand. How does this decision translate down the production chain and all the way to the end; to palletisation? Sometimes by changing a design and making it a little bit shorter and maybe a little bit wider, it becomes possible to optimize the design for maximum shipping efficiency.

The Cape Pack software offered by Esko, for example, incorporates value stream mapping that analyzes process steps and identifies waste along the workflow. Walmart uses a similar methodology to optimize shelf space and to optimize shipping utilization, enabling more goods to fit into its shipping containers. In other words, with the right software, sound process analysis, and giving the container shape some additional, strategic thought, it becomes possible to increase the amount of product in a container or on a pallet by five to 10 percent, thereby optimizing revenues and profits. See Figure 3.3.

As mentioned earlier, in addition to the aesthetics, shape can incorporate function, which brings in the

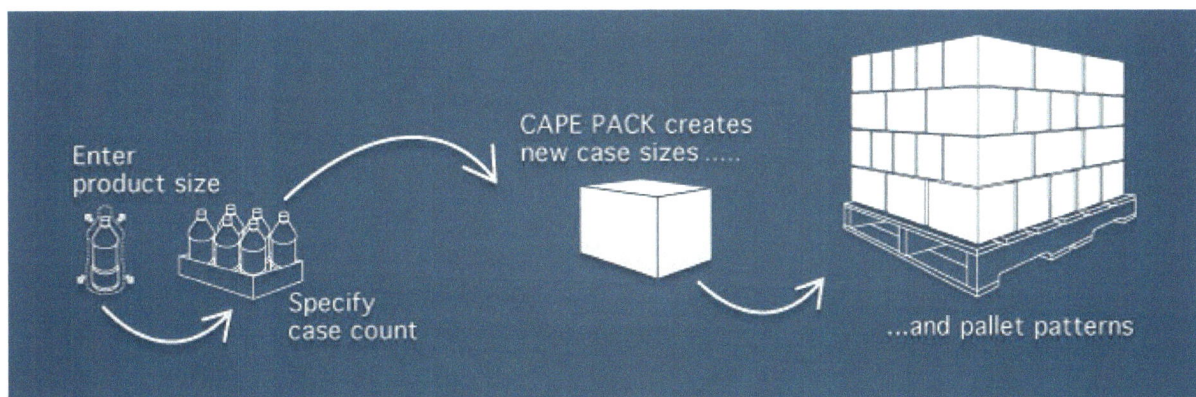

Figure 3.3 Cape Pack optimizes the shape to fit more products onto a pallet and shipping containers. Source: Esko

Figure 3.4 (left) and Figure 3.5 (right) show examples of how feel and touch can be built into the container shape and graphics. Source: Esko

role of ergonomics in pack design. Figures 3.4 and 3.5 show two examples of this with the functionality of grip and the sense of touch coming into play. The value of tactile appeal and the ability to evoke positive associations through the sense of touch are important and value-adding dimensions to consider in your container shape and design. Because shrink sleeving provides a 360-degree canvas, some of that area can be used to incorporate tamper evidence; this and other functional design elements like child safety features add value to the package design, but shrink sleeve labels all come down to graphics – picking graphics that best suit and follow the container shape, and then pre-distorting the graphics to fit.

With today's state-of-the-art software, like the shrink sleeve plugin for Adobe Illustrator from Esko, the software will pre-distort the design to maintain perfect consistency after heat-tunnel distortion. State-of-the-art pre-press tools will support the production process and make sure the design maintains its integrity wherever images, logos, text or barcodes are placed on the 360-degree canvas.

As previously mentioned, the early stages of the shrink sleeve design process need to bring together the brand owner, the substrate supplier, a structural designer and a graphic designer. The message should be, 'Think before you shrink'. Think

holistically about the container to avoid any unwelcome surprises.

SHRINK SLEEVE PRE-PRESS WORKFLOW
There are many software solutions available today that can simplify the shrink sleeve workflow process, but, in general, pre-press consists of three steps (see Figure 3.6). The first step is to prepare the graphics. For that purpose, Esko has a software suite that conveniently plugs in to Adobe Illustrator and is intuitive to use and relatively easy to learn.

Reviewing Figure 3.6, you will see that the lower left hand corner shows a file format called COLLADA. COLLADA is an XML-based three-dimensional asset exchange schema, and the acronym stands for 'Collaborative Design Activity'. It is an industry standard file format to exchange three-dimensional design files, just as PDF is for graphics data. Invented by Sony over a decade ago, this open source file format is widely used in the industry today. Any files with the filename extension '.dae' (digital asset exchange) or '.zae' (zipped asset exchange) can be exploited in three-dimensional design applications like Maya, Rhino, MODO, and Cinema 4D. All of these high-end three-dimensional design applications can generate assets in the COLLADA file format.

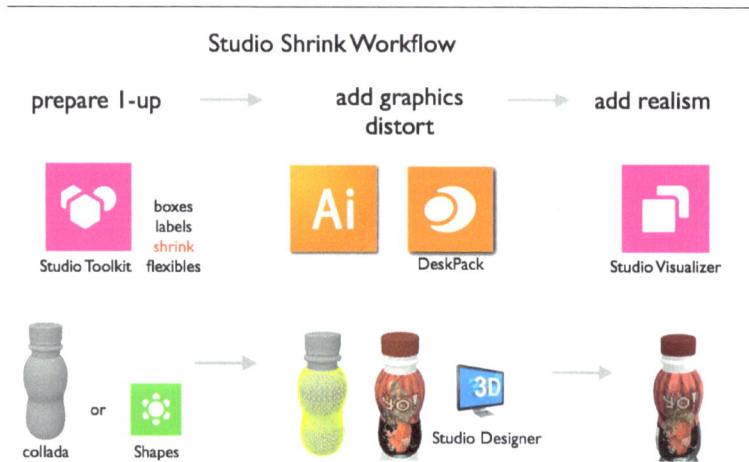

Figure 3.6 The shrink sleeve design, origination and pre-press process. Source: Esko

Once we have created our primary container shape in the COLLADA file format, the next step is to add graphics and prepare for the sleeve pre-distortion. Esko has a software application called 'Studio' for this purpose. This is a standalone software application on Mac or PC where a virtual sleeve is applied onto the primary container. This is where the typology, the three-dimensional landscape, can be determined.

Think of it as a landscape. That intelligence is now used to pre-distort the graphic design in such a way that, after shrinking the printed sleeve in a shrink tunnel, the design looks just as it was originally intended.

At this point, you can create a virtual pack shot of your shrink sleeve design, even before producing a physical mock-up or starting production. This visualization is a key component in the review and approval process with the brand owner. As this step comes early in the design process and all design objects are dynamically stored in standard formats, it is still very easy to modify the design of both the primary container and the graphics design without accruing any manufacturing costs whatsoever.

The usefulness of these virtual pack shots goes beyond the review and approval process. Once the package is approved and ready to go to production, these pack shots can be used in the go-to-market process, being featured on e-commerce web shops or appearing in social media campaigns long before the first dollar is spent on actual manufacturing.

The exchange of virtual pack shot in the review and approval process happens again with a standard file format, this time using standard PDF. The PDF file format has the ability to incorporate three-dimensional data. Acrobat Reader, the free version, can visualize this three-dimensional information on screen, enabling the operator to review all 360-degrees of the container, just like any other stakeholder in the supply chain can.

The disadvantage of using standard PDF is that not every aspect of the finished container can be visualized. Finishing and embellishments like coatings, varnishes, foils and embossing are all visual effects that materialize when interacting with the light sources in a certain location, e.g., in a retail environment. As

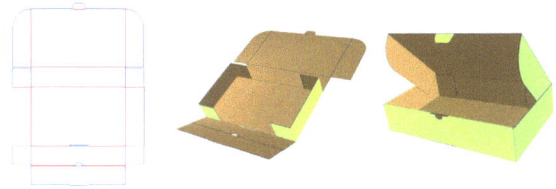

Figure 3.7 Creating a dieline for a box or folding carton. Source: Esko

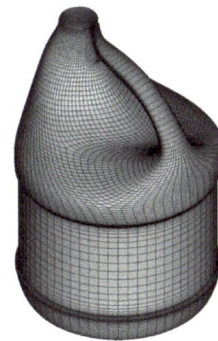

Figure 3.8 The complex die line for a container. Source: Esko

such, Acrobat cannot show the light casting off the substrate on to a surface or show the shininess or transparency of the substrate.

To close this gap and enable virtual evaluation of material and finishing effects, Esko has brought to market another plugin for Adobe Illustrator called 'Studio Visualiser'. With this tool, there is not only an on-screen, realistic visualization of a primary pack with the sleeve applied, but there is also the option to output high-resolution pack shots. The software supports a choice of lighting environments for a more

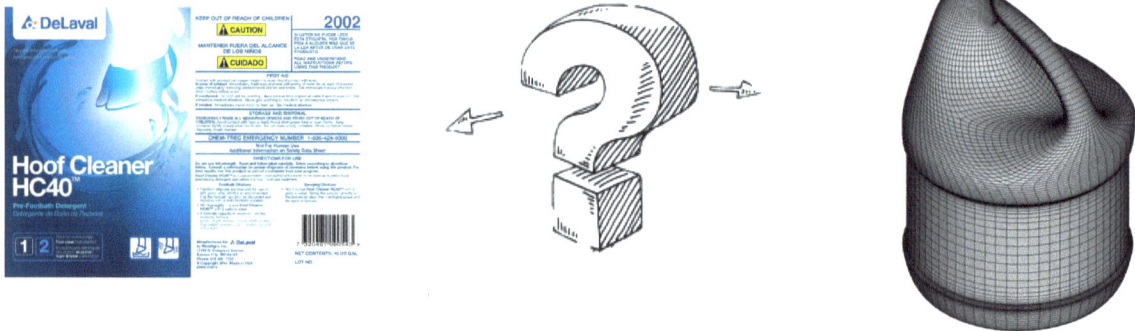

Figure 3.9 Regardless of the shape of the container, the design will be created on a two-dimensional rectangular space. Source: Esko

realistic pack shot and for product movies. Effectively, in the review and approval process, this software tool enables communication of hyper-realistic bottles in a store environment, typically done over a digital movie file format like .mov or .mp4. These file formats are viewable on mobile devices, meaning that the review and approval process is no longer limited to the confined space of an office.

CREATING CAD DATA AND SUPPLYING ARTWORK

What has been described so far is finding the balance between what is required and what is desired. In essence, the starting point is CAD data: three-dimensional data describing the primary package, be it a box, a folding carton, a display or a bottle. In paperboard packaging, the CAD data are easy to see: take a box, unfold it and what you end up with is the die-line, which can be seen on the left hand side of Figure 3.7.

The CAD data for a container and the die-line for a three-dimensional design, however, are much more complex. Moreover, it is not possible to cut open a canister and unfold it. A typology needs to be followed to create CAD data such as the one shown in Figure 3.8.

Sometime the CAD data for such containers

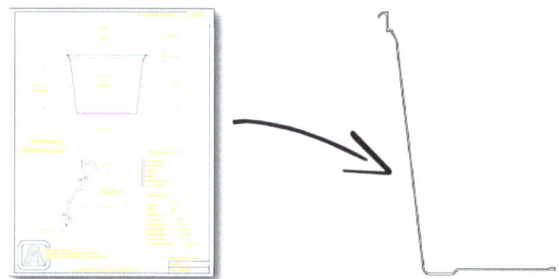

Figure 3.10 Creating a profile from a drawing or catalogue. Source: Esko

originate from the container manufacturer or from the brand owner or agency. There will be specification sheets for the containers, whether bottles, jars or tubs; sometimes it can be created from the outline of the container with the measurements; sometimes it can be created from a CAD drawing; sometimes it can be created by scanning. Each of these will be discussed in the next few paragraphs. The challenge (Figure 3.9) is to take a flat graphic design and be able to position it on the three-dimensional container

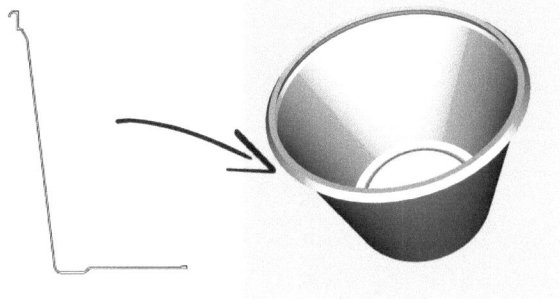

Figure 3.11 Creation of a three-dimensional container using Toolkit for Labels. Source: Esko

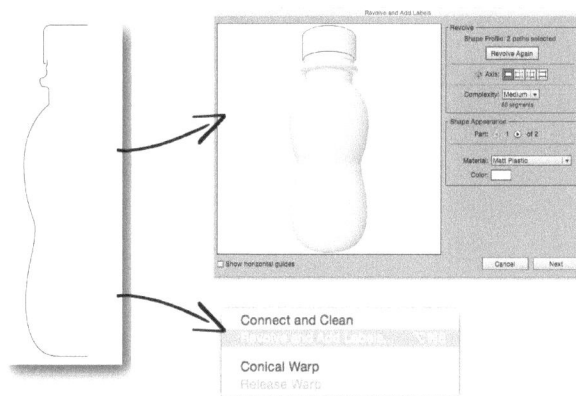

Figure 3.12 Using Esko Toolkit for Labels, a profile can be revolved into a shape. Source: Esko

die-line, ultimately creating the three-dimensional decorated container image. Regardless of how the container die-line is created, it will always require a file and physical samples. In other words, the three-dimensional graphic design will always be created on a two-dimensional rectangular canvas.

Apart from the container manufacturer's CAD data, the starting point may be a two-dimensional illustration, as shown in Figure 3.10. On the left is a drawing from the tub manufacturer. From this shape, we can create an outline in Illustrator, thereby giving us a container profile. It is even possible to take a picture of the container and upload it into Illustrator to outline the container profile.

Once traced, a toolkit called 'Toolkit for Labels' allows the profile tracing to revolve 360-degrees, creating a three-dimensional design in just seconds (Figure 3.11). It is not necessary to know all of the high-end design programs, but the disadvantage of this is that it only works for symmetrical products like, for example, a water bottle or anything that can revolve around the Z axis.

Figure 3.12 shows a profile shape in Adobe Illustrator. Select the 'revolve' option and a window will pop up to set the axis, indicate which material is going to be used – matte plastic, glass – and allow for color to be added.

Next, this information will be incorporated into the three-dimensional model to bring the concept to life.

Non-symmetrical shapes present a greater challenge than do symmetrical ones, but we can rely on sophisticated software to handle these types of containers as well. Just as Apple offers music, videos and applications in the iTunes store for its users, Esko offers countless container shapes in its Shapes Store, which are available for download. Figure 3.13 shows just one page (of almost 20 and growing) in the Shapes Store.

A fixed library of primary container CAD data has its limitations, however. One such limitation is that the designs are not 'parametric', meaning that if, for example, we download a European glass jar, the jar may have a 250g capacity. But what if we need a 300g or 500g jar? In fact, there is a way that we can work around this. Simply download the 250g shape from the Shapes Store, which runs in Illustrator, and select 'File', 'Open', from Shapes Store. Download the shape and bring that into MODO, whose logo you'll see in the lower left hand corner. Once in MODO, simply change the container specifications. The biggest challenge is understanding how to bring an object from the Shapes Store into MODO and change it: the best way to do so is by extruding it in a Z axis or make it more elongated in the X or Y axis.

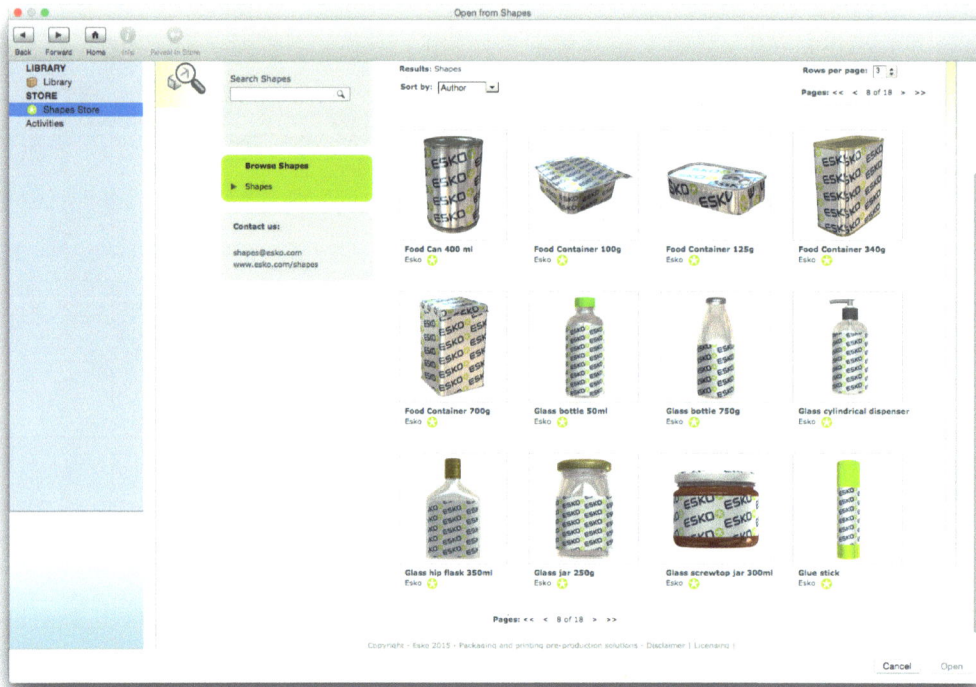

Figure 3.13 A whole range of shapes can be downloaded from the Shapes Store. Source: Esko

The last element to discuss pertains to odd shapes that cannot originate from a revolution or from a library download. Three-dimensional scanners, such as the one pictured in Figure 3.14, can handle these odd shapes with ease. At a cost of roughly $3,000 to $4,000, they are a relatively low-cost investment. To use, the odd-shaped container must be first sprayed with a white opaque ink. The container is then placed on the Lazy Susan that rotates very slowly as an eye in the scanner scans the container. Once the object is fully scanned, which can take up to an hour in high resolution, it can be uploaded into one of the software components to be cleaned and optimized.

Additionally, some websites like Turbo Squid offer designs for $30 to $40, all of which are uploaded by a network of designers. You can download these designs and customize as needed.

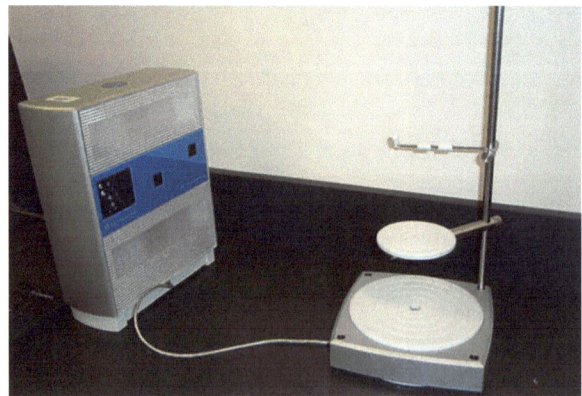

Figure 3.14 An example of a three-dimensional scanner. Source: Esko

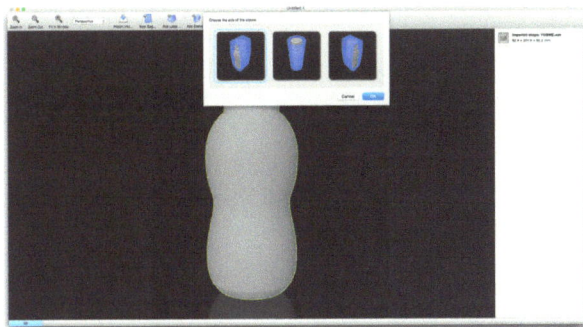

Figure 3.15 With the container shape created, the next step is to determine how the sleeve will be applied to the container. Source: Esko

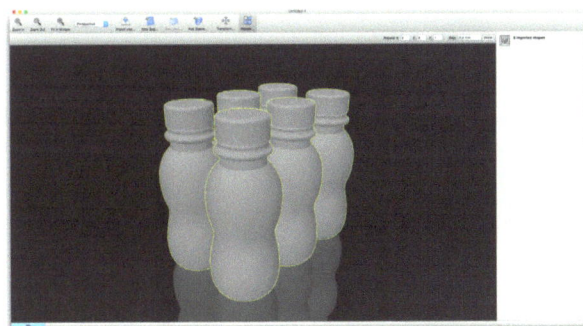

Figure 3.16 Step and repeat the container shape to create a six- or 12-pack. Source: Esko

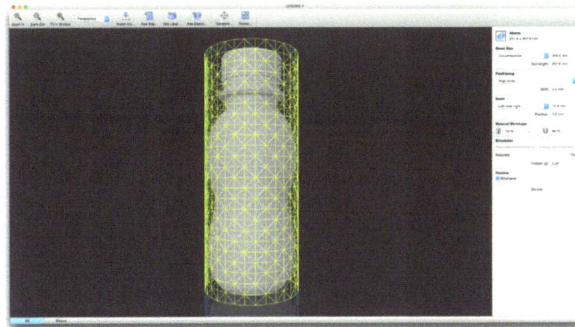

Figure 3.17 Specifying the production parameters. Source: Esko

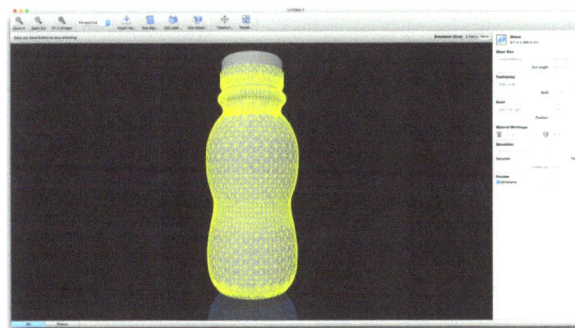

Figure 3.18 The sleeve can be saved and measurements can be provided for the graphic designer as they begin the process of adding graphics to the sleeve. Source: Esko

APPLYING THE SHRINK SLEEVE TO THE CONTAINER

With the three-dimensional container created, we will now turn our attention to applying the sleeve to our container, which we do in Toolkit (Figure 3.15). When we open the file (in either OBG or COLLADA file format), we determine the best method for applying the label to the container using the X, Y and Z axes. We can also step and repeat the container shape to create, for instance, a six-pack or a 12-pack of the container (see Figure 3.16).

Once we know how the sleeve needs to wrap around the container, the next step is to specify the production parameters, i.e., layflat, slit width, positioning, seam and shrink parameters, all of which are shown on the right hand side of Figure 3.17. The first item listed at the top of the panel on Figure 3.17 is the circumference, which gives the following options: layflat width, which is the circumference of the bottle without the seam, or layflat with the seam. If the packaging buyer or the creative agency provides

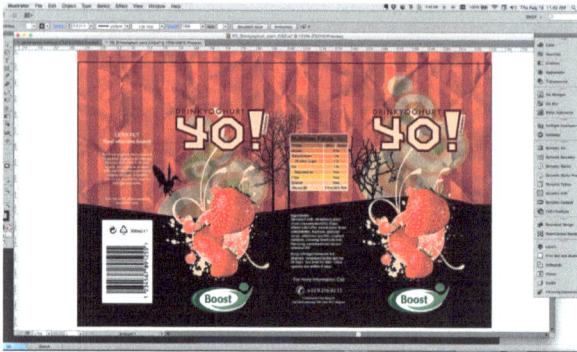

Figure 3.19 The sleeve graphics in Adobe Illustrator. Source: Esko

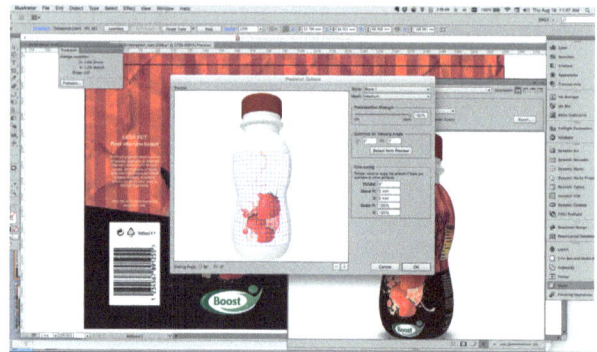

Figure 3.21 Artwork, text and images can easily be distorted using the Esko Studio for Shrink plugin. Source: Esko

Figure 3.20 The container structure and graphics come together for the first time. Source: Esko

Figure 3.22 An image in Illustrator of what the decorated sleeve looks like. Source: Esko

these dimensions, they can be verified at this stage.

Positioning of the seam is the next step of the process, with the key question asked being, 'Where should the seam be positioned?'. For a 360-degree container, this is perhaps an inconsequential question, but for a detergent or window cleaner trigger bottle container, the question bears greater importance. Generally, we opt to place the seam on the side of the container adjacent to where the wrist engages with the bottle and away from the graphics. At this

stage, we have the ability to visualize the seam in the software; we see it as a line. The software also allows us to enter a different view mode where we can position and turn the seam to the precise location we want it in a way that it does not interfere with the label graphics.

The final step involves inputting the material shrinkage parameters, both in the machine direction and the transverse direction, into the software, taking

Figure 3.23 The software allows for the selection of different lighting environments. Source: Esko

Figure 3.24 Use the window to select one of the many substrate options available. Source: Esko

into consideration the effects of the heat tunnel. Shrinkage always depends on the material being used, of course, and the stiffness of the material, not to mention a friction factor that takes into consideration how easily the sleeve slides over the substrate.

When we enter these variables into the software, what follows is a truly interactive process. Click on the 'Start' button and the process unfolds before the

operator's eyes. The simulation shows in real time how the CAD data of the shrink sleeve conforming to the container. The operator also has the ability to review the process at varying speeds.

Once rendered, it is now possible to preview the layflat of the sleeve and save the CAD data of the sleeve (Figure 3.18). It is also possible to toggle between the layflat in two-dimensional and three-dimensional formats. When in two-dimensional format, we have the exact label measurement readouts, which the graphic designer and illustrator will need as they add graphics to the sleeve. They will then know precisely how to position the graphics and which areas to avoid, e.g., the seam area.

In Figure 3.19 is a visual depiction of the flat sleeve graphic in Adobe Illustrator. We will open this file in the File menu and review the recently-created sleeve, which is in COLLADA format. What happens next using the Designer Studio software within Adobe Illustrator is quite incredible: we see in Figure 3.20 a preview of the labeled container. This is the decisive moment where structure and graphics come together for the first time in the design process. The beauty of this is that we see how the structure and graphics converge before placing a single container into a heat tunnel, which results in significant efficiencies and cost savings to the process.

Proper distortion of the sleeve artwork is possibly the most difficult aspect of shrink sleeve label design, and the amount of shrink necessary depends on the shape of the container. If we review Figure 3.21, we can see that the label design will require significant distortion in the center and towards the top of the container; in other words, we observe distortion in the Z-axis from the top to the bottom of the container.

Let's take a look at the graphic of the strawberries in Figure 3.21. The Photoshop operator needs to select specific graphic elements and look at the crop size of the strawberry to achieve the perfect distortion. The operator can continue by selecting other elements on the container and working in a similar fashion.

Figure 3.22 provides us with a visual of what the finished product will look like, and it takes into account the different styles of distortion, which pertain to the angle from which we view the sleeved

Figure 3.25 Export the COLLADA file and save to Esko WebCenter for further review. Source: Esko

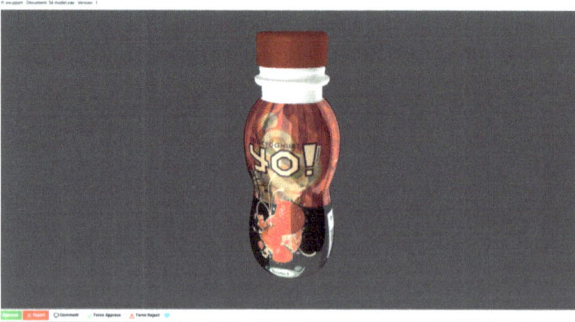

Figure 3.26 The file can be approved in WebCenter using any browser or mobile platform. Source: Esko

container. For instance, if the container is looked at straight on and is then rotated and viewed from the side, is the graphic going to follow the shape? How will the graphic appear from varying angles? Again, the software allows us to take these perspectives into consideration and to fine-tune for these nuances.

Another option within the software is to accommodate for different lighting environments. Looking at the bottom right window in Figure 3.23, it

is possible to select from a number of lighting environments – from a supermarket aisle to an airport, a warehouse or outdoors. Each option offers its own lighting environment, and each option will provide a different type of reflection on the shrink sleeve material. It is even possible to stage the product in a virtual retail environment where the operator can walk through the aisle, select the product (or any other product on the shelf, for that matter), look at it, and place it back on the shelf. All of this is possible with Store Visualizer, a software solution that offers Virtual Reality (VR) helmets and touch screen support, taking this experience to an entirely new level.

Taking this simulation process even further, Adobe Illustrator allows us the option of selecting different label substrates to understand, for instance, what the design looks like on a PVC film versus a transparent film. A list of standard materials, much longer than can be seen on the window in Figure 3.24, is available for selection, and the results can be communicated to the customer for approval.

We can accomplish this by exporting a COLLADA file that includes graphics (Figure 3.25). When this file is uploaded to a web-based collaboration solution, like Esko WebCenter, for example, all stakeholders have the ability to review the final design together virtually, saving on shipping costs and reducing the production process costs.

With the project managed in WebCenter, stakeholders have the ability to create multiple iterations and review many versions until everyone is satisfied.

Figure 3.26 shows an example of how interactive tools like Studio integrate with a web-based collaboration and approval solution called WebCenter. All operations happen from within Adobe Illustrator, but in the background, WebCenter will run automated workflows to ensure that external partners are invited to view the design in the browser of their choice. Invitees can even review designs via their mobile device or tablet.

Studio can also combine and manipulate different shapes together, such as we see in Figure 3.27. This example combines a carton with three containers, which represent three different flavors of honey in one cardboard container. We typically see combinations

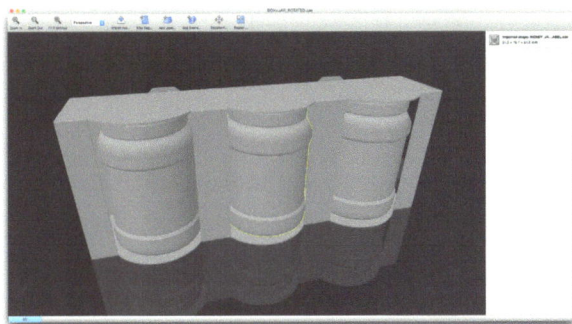

Figure 3.27 Esko Studio Toolkit combines different shapes. Source: Esko

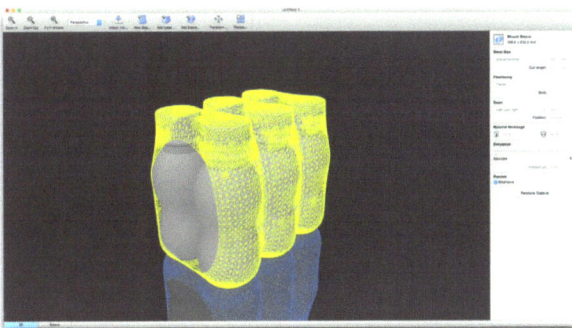

Figure 3.29 A finished shrink sleeve multi-pack of dog food cans. Source: Esko

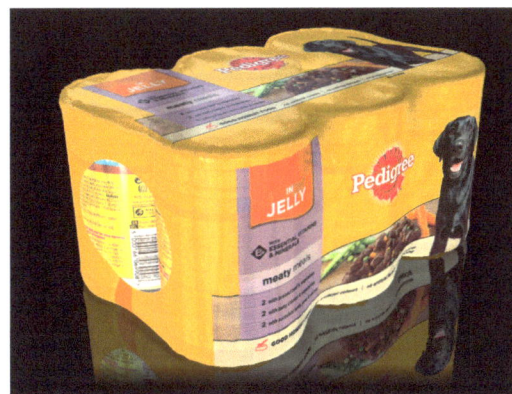

Figure 3.28 Applying a shrink sleeve design to a six pack. Source: Esko

like these when the brand owner launches promotional campaigns for the product. This mock-up can be created by bringing the components together in Studio.

Figure 3.28 shows an example of a sleeve being created for a six-pack. This type of sleeve will take a bit longer to render, though, given the complexity of the calculations.

A finished multi-pack of dog food cans is shown in Figure 3.29. The label going around the cans was applied in Adobe Illustrator. Step one involved measuring the six cans in this multi-unit pack. Step two involved applying a shrink sleeve wrap around these six measured cans. Step three involved applying graphics to the shrink sleeve.

BARCODE CONSIDERATIONS

How are barcodes included in the design of the shrink sleeve label, and what factors must we consider when applying them and determining their positioning on the label? Esko offers tools to apply the barcode within Adobe Illustrator. Supported barcodes include all standard barcodes, QR codes, UPC codes and smart barcodes.

What considerations need to be taken into account when positioning barcodes? First, place the barcode on the least curved position of the container; it is necessary to make sure that the barcode is readable. Electronic barcode readers can determine whether a barcode is readable, and barcodes can even be graded for their readability. To do so, a more robust solution called Automation Engine is required. Automation Engine is powered by GlobalVision, a set of tools for the automated inspection of barcodes, braille text, and spelling. A spelling checker is available to spell-check designs in background, as part of an automated workflow.

Second, position the barcode vertically rather than

Figure 3.30 Barcode positioning considerations. Source: Esko

BARCODE CONSIDERATIONS

- Place barcode on the least curved portion of the container.

- Position the code vertically instead of horizontally.

- Do not place the code too close to the seam, which can affect scanning ability.

horizontally on the label. In doing so, the barcode becomes less susceptible to shrinkage distortions that may render it unreadable.

Third, do not position the barcode too closely to the seam. Doing so could affect readability as the risk is present of clipping part of the barcode off. These points, which are summarized in Figure 3.30, all seem fairly intuitive, but never assume that all stakeholders follow these rules. Designers sometimes do not work closely with the sleeve converter, or the revision process is not always as thorough as can be. In other words, things can – and do – go wrong.

To summarize, the brand owner, the converter and the graphic designer must all work in tandem to make this complex and nuanced process work, and to deliver a perfect shrink sleeve label whose design aligns perfectly to the contours of the container it is decorating. Fortunately for all of these stakeholders, we have seen advances in software that allow us to not only create a beautifully-designed shrink sleeve label, but also allow us to simulate the transition from a two-dimensional design to a three-dimensional work of art that captures the consumer's attention and compels them to purchase the product, which is the reason we opted for shrink sleeve labeling in the first place!

Chapter 4

Printing technologies and inks for shrink sleeve decoration

Inks and the printing process are important considerations in the shrink sleeve production process. Shrink sleeves help the consumer to make a choice; they differentiate a product on the shelf with 360-degree decoration, and graphics and color play a significant role in the success of this labeling technology. Ink formulation and selection are therefore critical inputs to the finished shrink sleeve. They are also challenging to get right, given the very unusual demands that the shrink sleeve process puts on ink – that is, adhere to a film, and then shrink along with that film in a heat tunnel!

That being said, what do we need to know about printing processes, inks and coatings to make informed selections for optimum shrink sleeve printing and performance?

Sleeves can, of course, be printed with virtually any of the printing processes. While rotogravure remains the most popular method for high-volume sleeve printing, there has been significant growth in the use of flexo, and digital has grown in popularity for shorter runs.

While the majority of shrink sleeves are printed on mid- to wide-web presses, there has also been significant growth in sleeve printing using the narrow-web process, especially using UV flexo, and to a lesser extent, water-based flexo. Whatever the sleeve printing process, and whether it is wide- or narrow-

web, it is important to have an understanding of the inks used for sleeves. This requires us to look at the ink chemistries and the different types of raw materials used.

SHRINK SLEEVE INK CHEMISTRIES

The primary ingredients in inks transcend the different printing technologies. Every ink has a pigment, a resin, a diluent or a solvent, as well as various additives that are incorporated to improve performance. These ingredients are all summarized in Figure 4.1, and each will be examined in greater detail.

Pigments. The first and most important aspect of ink to consider is the pigment, the coloring matter that goes into the ink to give it, say, a red, blue or green color. The different pigments used in shrink sleeve

INGREDIENT	SOLVENT	WATER-BASED	UV CURING	OIL / OFFSET
PIGMENTS	Yes	Yes	Yes	Yes
RESINS	Nitro	Acrylic	Oligomers	Phenolic Alkyd
DILUENTS	Solvents	Water/ amine	Monomers	Mineral / Vegetable Oil
SOLVENTS	> 30%	< 5%	~0	Zero
ADDITIVES	Wax Antifoam Silicones Plasticizers	Wax Antifoam	Wax Photo Initiator Stabilizers	Wax Stabilizers Fillers

Figure 4.1 The generic raw material used in the different ink types. Source: Flint Group

Pigments	
Physically & chemically stable Non-soluble	
Denoted by ColourIndex number (CI#):	**Red 57.1** **Red 184** **Red 177**
CI# indicates properties:	**Hue, fastness, cost Important to know CI#**
Heavy metals content:	**CONEG, RCRA, F-963/EN-71 Barium, mercury, hex. chromium, lead, cadmium, antimony, arsenic and selenium primarily**

Figure 4.2 Pigments used as raw materials in sleeve inks. Source: Flint Group

printing inks may appear the same color when first printed, but they can perform quite differently after they are processed in any way – for instance, when they have passed through a heated shrink tunnel, are filled or used in a chemical environment, or are exposed to strong sunlight. In these examples, the pigments may change color or 'bleed' to differing degrees. An understanding of pigment chemistries and how they are defined can therefore be an important factor in ink sourcing.

Pigments are identified, globally, by a Colour Index™ (CI) number that defines the particular chemistry of the pigment, enabling it to be classified along with other products whose essential colorants are of the same or different chemical constitutions. For example, Figure 4.2 shows the CI numbers for Red 57.1, Red 184 and Red 177. These three different Red CI numbers will each give a magenta or rubine shade, but the chemistry of each of the pigments is different: differences that can affect performance in terms of light fastness, chemical resistance or heat fastness. It is worthy of note that the use of incorrect pigments in shrink ink can result

in the ink changing color upon shrinking in the heat tunnel.

Chemical resistance of the chosen pigment is also important. For example, in the case of a household chemical where the container is being filled after sleeving, it is imperative that the ink be resistant to the chemicals to avoid bleeding, should the chemical make its way between the sleeve and the container.

Consideration and understanding of ink pigments, and the end-user requirements of the ink, are always important. Take, for example, the requirement for light fastness. The four yellow colorants shown on the chart in Figure 4.3 are all Pantone® yellow. If they were printed on a sheet of white paper, they would all appear as Pantone® yellow. Place them in a light fastness tester or expose them outside to light, however, and the top one, which is a dye-based material, will fade to the point of disappearing after a 24-hour time period. Again, depending on the choice

LIGHT FASTNESS CHART

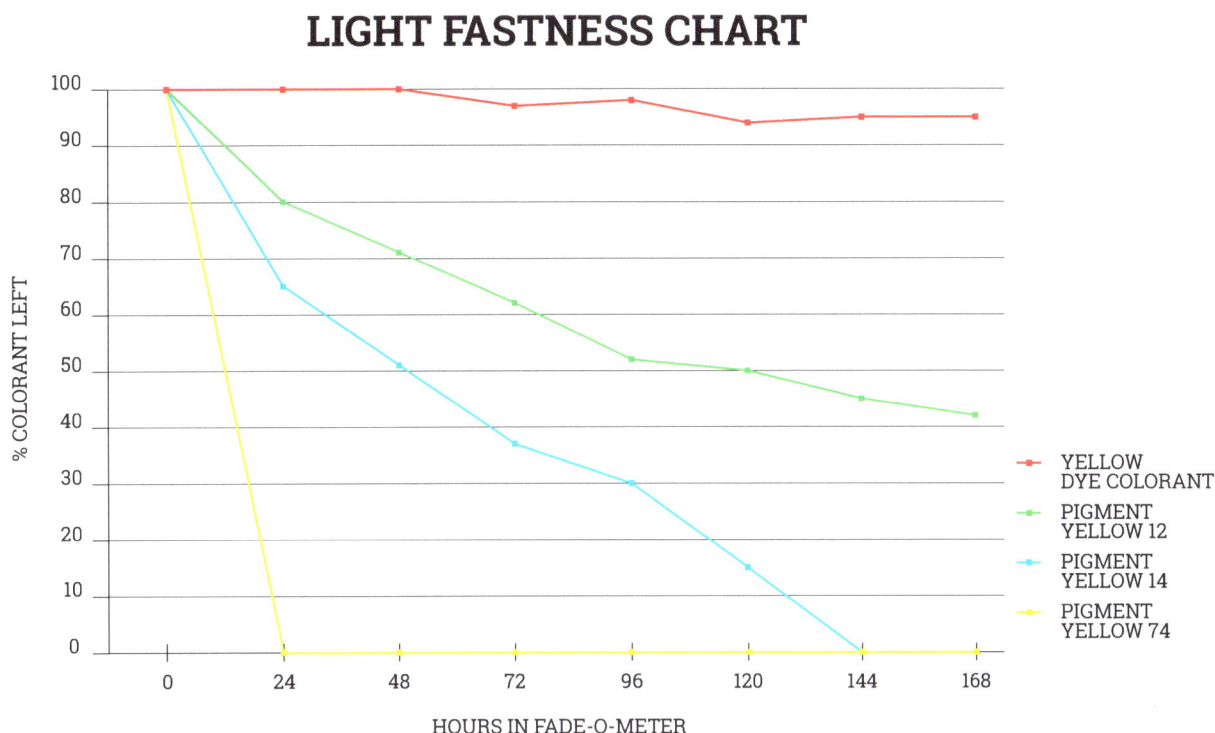

Figure 4.3 Light fastness chart for different yellow pigments. Source: Flint Group

of pigments, degradation of the color will occur over time with exposure to UV light.

It is also possible to select pigments with chemistries that hardly degrade at all. Despite these pigments being more expensive, they may be an absolute requirement, depending on the application in question. For example, if the sleeves are for a chemical that will be stored in a greenhouse or outside at a hardware store, then the ink's light fastness is a requirement that cannot be compromised. The sleeve printer needs to make sure that the pigments in the inks are appropriate for the application.

Resins. Resin chemistry forms the backbone of any ink, and it is with resins that pigment manufacturers, ink suppliers, and chemists select from a range of options. Some of the most common are:

- Nitrocellulose
- Polyamide
- Acrylic
- Phenolic
- Epoxy Acrylate – used in UV/EB inks and coatings
- Urethane, Polyester – used in solvent and UV/EB inks and coatings

Resins provide the main basis of the ink and will affect adhesion, flexibility, resistance properties, drying speed/cure, and overall end performance of the printed material.

Nitrocellulose, urethane and polyamides are commonly used in solvent systems. Acrylics can be used in water-based systems, and there are different epoxy acrylate, polyester acrylate, and urethane acrylate chemistries that can be used in radiation curable systems.

Each chemistry provides a different performance. With shrink sleeves, the most fundamental requirements of the ink are that it provide good adhesion, that it be flexible, and that it can follow the shrink profile of the film in the shrink tunnel.

Other properties that ink suppliers consider are resistance properties, running speed, and ink drying-speed requirements. Ink suppliers will select different chemistries based on these requirements to optimize performance.

Diluents. Another raw material component of inks are diluents. These are used to reduce the viscosity of the ink. For water-based ink, the diluent is primarily water, and for solvent-based ink, the diluent can be alcohol/acetate blend. UV systems use monomers. Diluents are always selected with the shrink application in mind. As mentioned in an earlier chapter, with solvent rotogravure systems, it is not possible to use ethylacetate as a diluent as it impacts the chemistry of the shrink film. Since every solvent system has its preferred solvent, should that solvent be unusable with shrink film, then it follows that not all solvent systems can be used with shrink film.

Additives. There is a wide range of additives added to inks to achieve various performance characteristics. Among the most common are: defoamers (or anti-foaming agents), waxes and silicones, matting agents, photo-initiators, adhesion promoters, surfactants, and optical brighteners. Among the most important additives for shrink sleeve applications are waxes and silicones that are added to achieve the appropriate coefficient of friction on the last-down white or slip coating.

From the ink manufacturer's perspective, corona treatment of shrink films by the printer is important. A light bump treatment of almost any film with a corona treater – whether shrink labels, a flexible packaging construction, or a pressure-sensitive film label – is always recommended. As films age, they will lose treatment. As film is handled, or as it sits in the warehouse – with temperature and humidity variations – the treatment level will diminish. A light bump treatment will bring the dyne level back to where it was intended to be. Shrink films are very printable in terms of print performance, but achieving ink adhesion is not always so easy. PVC and OPS are

generally very good for adhesion, while PETg is not as good. Polyolefin can also be challenging for some chemistries, such as water-based or UV.

Solvent chemistries, whether for rotogravure or flexo, generally offer the best adhesion to shrink films. With water-based, radiation curable UV systems, and digital, films will generally need to be treated. In some cases, it may be necessary to use a primer.

If manufacturers recommend treatment, be careful not to over-treat films because this can create issues of blocking in the roll. Basically, corona treatment changes the surface characteristics of the film. The level of treatment is measured in dynes. The printer can undertake some easy dyne tests to make sure that the right dyne levels are being achieved. Corona treating puts a carboxyl or hydroxyl functionality on the film surface so that the ink chemistry can chemically bond to that film surface. With the proper resin selection, the right chemistry in the ink and the proper application and drying or curing, it should be possible to adhere to these materials. Nevertheless, a light treatment during the printing process is generally recommended.

CHALLENGES FOR SHRINK SLEEVE INKS

What are the market requirements, and what are the challenges for shrink sleeve inks? As Figure 4.4 illustrates, shrink sleeve inks require a very high color strength ink that can cure quickly and maximize press speeds. The presses run fast and inks are needed that can print, cure and dry quickly in order to be as productive as possible, achieve a good print quality, achieve good adhesion characteristics, and have a low odor. It is also necessary to meet various environmental regulations like Proposition 65, Nestlé and the Swiss Ordinance on printing inks, which has become the standard for food packaging inks in Europe, or FDA regulations in North America.

Ideally, the shrink profile of shrink film will not change when ink is applied to it. Nevertheless, a few layers of white or black ink coverage will surely start to change the shrink profile of the raw film stock - it likely will not shrink as far or as fast. Ink systems are challenged to be as robust as possible to adhere to the variety of films available.

Inks also need to exhibit good surface slip

Challenges for sleeve inks

Very high color strength, high cure speed, excellent press and print performance, low odor, and meet Environment, Health and Safety (EHS) regulations.

Technical challenge to maintain ability to shrink,as ink is cross-linked when drying/curing.

Adhesion to wide range of substrates (PETg, PVC, OPS, OPP, PLA, Polyolefin) without the need for primers.

Good surface slip properties; use for high speed seamers and applicators.

Hazing of last down, low coefficient of friction (COF) over-print varnishes.

Resistance to moisture, scratch, chemicals.

Opaque white is key! Higher opacity opaque white with excellent COF properties.

Minimize 'Sellotape' effect (which looks like 'wet T-shirt').

Figure 4.4 Summarizing the key challenges for shrink sleeve ink. Source: Flint Group

properties, especially for the seaming and application stages. The coefficient of friction (COF) is important for the last down white which, due to the sleeve being reverse printed, is the layer of ink that comes in contact with the tooling on seamers, applicators, and ultimately the surface of the container upon which the sleeve will be shrunk.

Hazing can be a challenge in situations where there is a clear, i.e., unprinted, area incorporated into a graphic design. While a coating may be desirable for COF reasons, the recommendation is generally to leave the window uncoated (or unprimed) to avoid the hazing that can occur. If the film must be primed or coated, then avoid using UV chemistry, which will get hazy as it shrinks. If it is necessary to put slip coatings or primers as a last down from a seaming standpoint, it is recommended to use a solvent-based or water-based chemistry.

Another phenomenon that frequently challenges ink in a shrink sleeve is called the 'wet look'. In fact, the 'wet look' can be divided into two totally unrelated effects. The first has become known as the 'wet T-shirt' effect, which is caused by moisture that gets trapped between the sleeve and the container as it passes through a steam tunnel. The 'wet T-shirt' effect is temporary and disappears as soon as the moisture evaporates. The second 'wet look' is more complex in its cause, and more complex to resolve. If a wet look develops, and it is not caused by moisture (as would certainly be the case if it was not shrunk in a steam tunnel, for example) then we refer to this as the 'Sellotape' effect. The 'Sellotape' effect is caused by the refraction of light when two very smooth surfaces come in contact with each other. In fact, it is observed when two panes of glass are put together, or, if the smooth inner surface of a shrink film is forced tightly against the smooth outer surface of a glass or PET container. See Chapter 7 for photographs of the 'Sellotape'effect.

The 'Sellotape' effect is really a nuisance as it diminishes the overall look and finish of the container. Fortunately, there are ways to overcome this issue. With certain ink formulations such as whites and clear coatings, the 'Sellotape' effect can be minimized or eliminated. These inks and coatings operate on the principle of creating a textured surface on the ink to create distance between, and eliminate the smoothness of, the two contact surfaces. This creation of space between the two contacting surfaces is generally enough to eliminate or greatly reduce the problem.

THE IMPORTANCE OF WHITE

As previously mentioned, most shrink sleeves are reverse printed. In other words, the ink is printed on the inside of the film, so the ink is trapped between the shrink film and the container. This means the

Static COF - spec ~ 0.4 – 0.5

is calculated from the force that is needed to start something that is lying still on the surface.

Dynamic/Kinetic COF - spec < 0.25

is calculated from the force that is needed to move something over the surface once it has started to move.

This value is always lower than the static COF. The dynamic COF is important for sleeves and should be low enough to make processing in seam machines easy and to make it easier to put the sleeves on the bottles.

If the static and dynamic COF are too low, it can be difficult to handle the web during seaming.

If the dynamic COF is too high, the sleeves will not work on high speed seaming and sleeving lines.

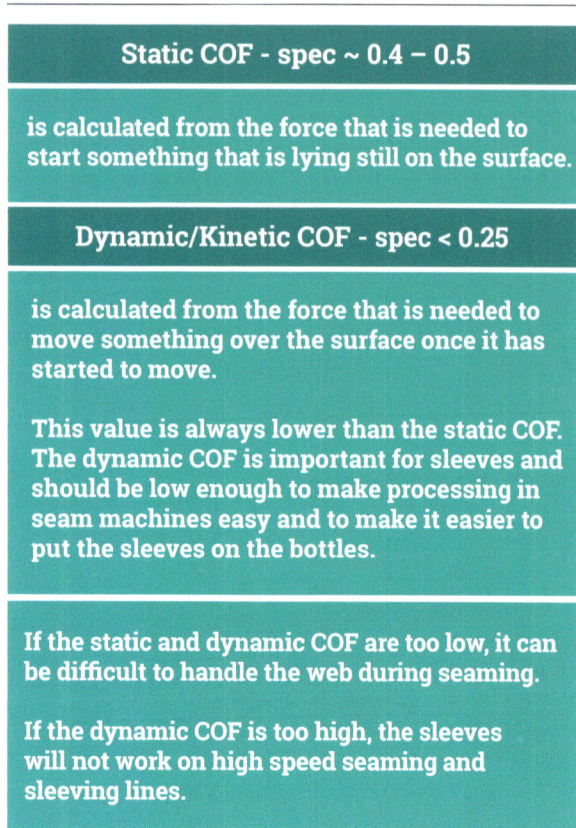

Figure 4.5 The importance of coefficient of friction (COF) with whites. Source: Flint Group

white is the last down – the colors are printed first, followed by the opaque white.

Most shrink sleeves have a lot of opaque white coverage, so the last down white is very critical. Coefficient of friction (COF) is a concern as the ability of the ink to slip over the tooling in high speed seamers and automatic application equipment is essential.

Sleeves generally call for high opacity. To achieve this, printers start putting whites down with a very coarse anilox, which becomes a problem. If whites are being put down with a 10 BCM (billion cubic micron) volume anilox (metric = 15.5 cm3/m2), then with high shrink applications the ink is actually going

to start piling up on itself and will create an effect referred to as 'tree barking' (see Chapter 7 for a photograph). What is generally recommended to achieve the desired opacity without laying down too great a volume is to do multiple bumps of white. Some printers will do two or three bumps of white with a 4-8 BCM anilox roller (metric = 6.2-12.4cm3/m2) to get the desired opacity.

Figure 4.5 outlines the recommended values for static and dynamic COF for shrink sleeves. The COF figures are critical and should be monitored very carefully. With a little practice, it is possible to develop a quick feel for what is acceptable by simply rubbing the inner surfaces of a sleeve between your fingers. You can easily feel when the two surfaces stick to each other, or when they are actually too slippery and hence hard to control on high-speed seaming equipment. It is highly recommended, however, to use a flatbed COF testing device in your testing process.

UV PRINTING OF SLEEVES

For the narrow-web printing of sleeves, UV is really a fast growing application in the market. However, there are some challenges with UV printing (Figure 4.6) that are worthy of discussion. Typical UV presses have mercury lamps, often referred to as arc lamps that generate a lot of heat. This means that a press with good heat management (i.e., chiller rollers, cool lamps, etc.) is required. Otherwise, the shrink film will effectively start to shrink on the press and create difficulties for print register as well as future steps in the process. Therefore, a challenge exists in using typical narrow-web legacy presses, as the heat needs to be a key consideration on these presses.

To overcome the problem of UV heat generation, some converters have turned to cool UV systems. These systems make use of reflectors and different types of light output and may require a change to the ink chemistry. A printer getting into shrink sleeves should talk to their ink suppliers about the type of UV system used on the press so that the right ink chemistry can be matched to the UV curing system; otherwise, the inks will not cure properly.

Another approach to managing heat on presses is to use chill rolls. Many converters will run the chill rolls very cold. Indeed, a lot of the press manufacturers will

Challenges with UV printing

Heat generation from UV lamps will distort films, especially thin films if there are not heat management techniques on press ('cool UV' systems, chill rollers, etc.)

If you do not have the proper ink formulated for 'cool UV' systems (including chill rollers, chill plates or cool UV lamps), then the UV inks may not cure as fast or cure as well, resulting in poor adhesion, poor moisture resistance, etc.

Often, the chill roller is too cool!

Proper curing for both surface cure and through cure (which will affect adhesion) can occur if applied with too coarse an anilox roller or at too fast of press speeds.

Figure 4.6 Challenges to consider when printing with UV curing. Source: Flint Group

suggest running the chill rolls at 18 degrees Centigrade (65 degrees Fahrenheit), when in fact this is too cold! They suggest running at this temperature because they are afraid that the films will distort if they do not. However, it should be noted that chemical energy slows down in a cooler environment. As it warms up, the rate of chemical reactivity increases. Therefore, when running presses with any UV or radiation curable technologies, look at the chill rolls and run them at a minimum of 80 degrees Fahrenheit, which will not cause the film harm or damage it in any way. Some films, especially PETg, will benefit from running the temperature closer to 100 degrees Fahrenheit.

At these higher temperatures, the film will be less brittle and will be less likely to break on press. Moreover, ink adhesion improves and curing speeds will be faster, too.

Converters prefer strong blacks and opaque whites, leading them to putting down inks with very coarse anilox rolls and, in some cases, getting to a point where the ink is too heavy and mercury light cannot cure through the thick film. Be careful not to fall into this trap, and work with ink suppliers to understand the proper applications. Determining which anilox roll line count and volumes should be used to print these inks will yield the strongest colors, the strongest densities, and the highest opacities. If you exceed these recommendations, curing problems and other issues will arise. When curing problems start, you will see that adhesion of the inks to the shrink films will be inadequate. Different formulas, especially for whites and blacks, can be applied to achieve higher density and opacity. It is advisable to speak with your ink suppliers about getting the right ink, with the right anilox, to maximize performance and speed.

UV LED SLEEVE PRINTING
Though still relatively new, UV LED is an ideal choice for shrink labels. This can be explained using the Wavelength Comparison chart in Figure 4.7. This compares the output of a mercury lamp – which comprises some 98 percent of the presses in the market today – with that of UV LED.

Reviewing the chart in Figure 4.7, we observe that the mercury lamp outputs of light in the short wavelength, or UVC range, is where ozone is generated. The infrared range, on the far right-hand side of the chart, is where heat is generated. Both ozone, an air pollutant, and heat are undesirable by-products of this technology. Excessive heat generated by the mercury lamp on a press is undesirable. While the web temperature should be warm, the heat generated by mercury lights – reaching temperatures of up to 300 degrees Celsius (572 degrees Fahrenheit) – becomes problematic.

With the UV LED wavelength, as represented by the line graph in the middle of the chart on Figure 4.7, we observe a very narrow wavelength of light primarily in the UVA range with a very high intensity output. Because UV LED does not emit light in the UVC or infrared ranges, no ozone or IR heat is generated from this technology. UV light does not utilize mercury,

Wavelength Comparison

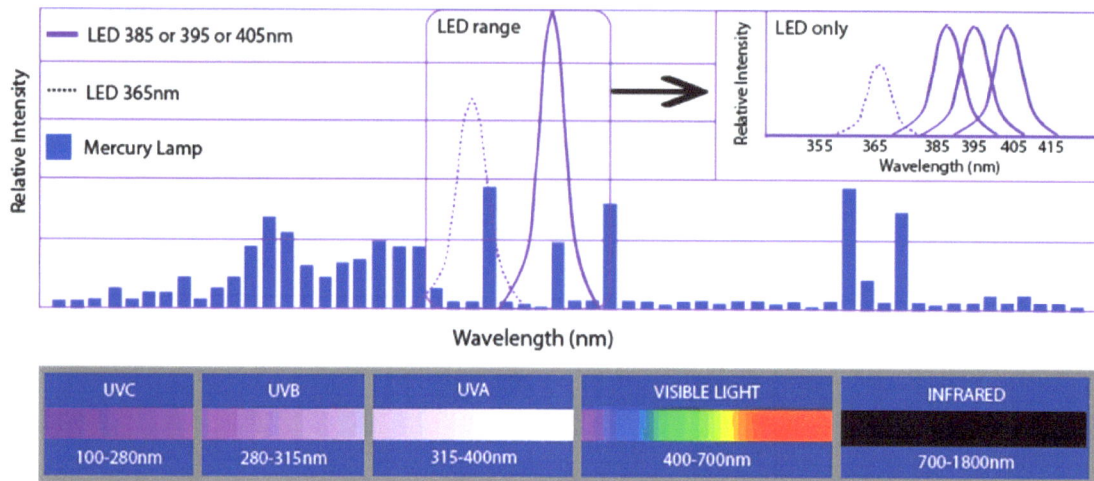

Figure 4.7 Comparison between Mercury and LED curing wavelengths. Source: Phoseon Technology

which is a toxin.

Since the light source of UV LED is quite different from that of the mercury light, UV LED technology requires different ink chemistry. It is important to note that the UV LED light source of UVA is a deep, penetrating light source. With a deeper penetrating light source and with higher intensity, this technology will cure inks that are darker, denser and more opaque at faster press speeds. When inks are cured better, they will adhere better, which will improve their performance in subsequent processing steps and for end user requirements on various applications. A central problem with traditional UV systems is that the UV mercury light does not cure inks very well, not to mention that lamps degrade over time and reflectors get dirty. While still a new technology to the industry, UV LED will be looked to more and more as a solution in the shrink sleeve industry and other applications such as pressure-sensitive labels and flexible packaging.

With digital UV inkjet printing, a process known as LED pinning may also be used. In LED pinning, the surface of the printed film is given a light bump of LED to just cure the surface of the printed ink film, but the ink layer is not fully cured. The surface is cured, ready to put down the next layers. Like inkjet, the film does not need to go through turn bars and so this approach works nicely. It would not work in a flexo operation because the film needs to go through a lot of turn bars, which means inter-station curing and full curing, just the same as it would be on a conventional press.

In general, LED is ideal for shrink because there are no concerns about managing the heat out of the press. The key points to remember are:

- Optimization of the UV curing process requires consideration of not only the UV light source, but also the ink chemistry being used.
- UV LED sources are equivalent to or better than existing arc lamp solutions, but require ink reformulation.
- The key to success is the relative higher power of UV LED and specialized ink formulation.
- UV LED is ideal for shrink sleeve applications!

UV flexo printing tips

It's all about CURE!

Proper anilox roller selection, ink selection (blacks and whites especially)

High opacity white requirements for shrink applications slow speeds as mercury light does not penetrate and cure inks that are highly opaque

Denser blacks do not cure at high speeds either,and often waste is generated when inks are found not to pass adhesion off press, or worse, downline

Proper press speed for lamps – depth cure

Maintenance of UV lamps, reflectors, etc.

Chill roller temperatures

Figure 4.8 Optimizing UV flexo sleeve printing inks. Source: Flint Group

Water-based flexo printing tips

Proper anilox roller selection and press speed

High air velocity drying (not too much heat!)

Catalyze inks (whites/last down) if going through steam tunnel

pH maintenance is critical!

Figure 4.9 Optimizing water-based flexo sleeve printing inks. Source: Flint Group

ACHIEVING THE BEST RESULTS

Having reviewed the requirements of the inks and ink formulations for shrink sleeve printing, we will now provide some tips and guidelines for obtaining the best results with each of the main printing processes.

UV flexo. In UV flexo printing, cure is paramount. To be successful in UV flexo printing (Figure 4.8), it is necessary to make sure the inks are cured well and that the right anilox roll is used with the right ink. Make sure that the press and the UV system, i.e., the lamps and reflectors, are well maintained. With high opacity whites and dense blacks, make sure the right ink products are used with the right anilox. Chill roll temperatures have a huge impact on conversion speeds, conversion success and performance of the substrate as well as inks. We recommend discussing

how to use optimal converting settings with the press supplier, or alternatively, the ink manufacturer.

Water-based Flexo. From a water-based standpoint, again, proper anilox selection is key. Talk to the ink suppliers and make sure that for the ink sets being used, the recommended anilox line count and volumes are correct. This will maximize press speeds and performance.

With water-based inks (Figure 4.9), it is important to not use heat to dry them. Rather, it is important to move as much air across the web as possible – air velocity is key. Converters often think that the best way to dry water-based inks is by using a lot of heat, which is incorrect. Additional heat is not required; just use air, but ensure high air velocity.

When using water-based inks for shrink sleeves, it is necessary to catalyze the last down product – which is usually a flood coat of white with a catalyst. This is necessary because there is a lot of moisture in the steam tunnel, and if that moisture contacts the inks, they will start to re-solubilize. Therefore, with water-based inks and steam tunnels, it is necessary to catalyse the last down ink.

Another critical factor with water-based inks is the pH level. Make sure to have a good maintenance program relating to pH.

Solvent based inks have already been mentioned, especially in relation to how acetates can

Figure 4.11 Examples of typical laboratory/press side tests. Source: Flint Group

---

UV/EB web offset printing tips

One side of shrink films generally is coated with antistatic coating.

If the label telling you which side of the film is coated is lost, curl the film into a tube and blowin it – the cloudy side is coated.

Be careful with ink/water balance.

Scumming is particularly problematic for seaming, which can leave residue on seaming felts, causing open seams.

If you print this side, the fountain solution can pick up the antistatic coating and become contaminated.

New fountain solutions are available that will not solubilize the antistatic coatings as easily.

Figure 4.10 Getting the best out of UV/EB sleeve printing inks. Source: Flint Group

---

damage shrink film. Make sure to buy the right inks for shrink applications. OPS is particularly sensitive to solvent attack. The other point is to ensure correct ink viscosity. Solvent ink suppliers will typically supply inks that are not press ready, as is the common procedure with narrow web. For wide-web presses, it is typical to provide a concentrate, which the printer dilutes to a certain printing viscosity. In this case, it is important to make sure to dilute the inks with the proper solvents and to the correct viscosity.

Web Offset. A common issue with offset printing is contamination to the seaming area from the fountain solution. This can be caused by the antistatic coating that is typically applied on one side of PETg film. Applied by the film manufacturers, this coating is designed to reduce static during the conversion process. While the general recommendation is to print on the non-antistatic side, some converters prefer printing on the antistatic side for reasons of adhesion and printability. Printing on the antistatic side is not recommended for offset application. See Figure 4.10 for an easy way to tell which side of the film is coated with antistatic, if it isn't labeled.

Also with offset, it is necessary to be careful with the ink-water balance to avoid ending up with scumming and other problems back in the fountain solution. These contaminants may transfer onto the web in the seaming area and create poor seams or even failing seams during the shrink process. There are new fountain solutions that are available that will not solubilize the antistatic and cause contamination in the seam area. This is very important when using offset printing.

Digital. Most of the digital inks will need a primer, and many film suppliers have pre-primed films for the digital applications, especially for HP presses. Or, the converter can buy their own primer and prime it in line. If the printer wishes to undertake their own priming, then they need to note that the seam area must be left clean and free of primer.

Primers and inks must be kept out of the seam areas. The seaming solution is a solvent, requiring raw film to raw film contact in order for the chemical weld to work.

TYPICAL PRESS-SIDE/LABORATORY INK TESTS

Ink manufacturers highly recommend that printers undertake some typical test procedures (Figure 4.11)

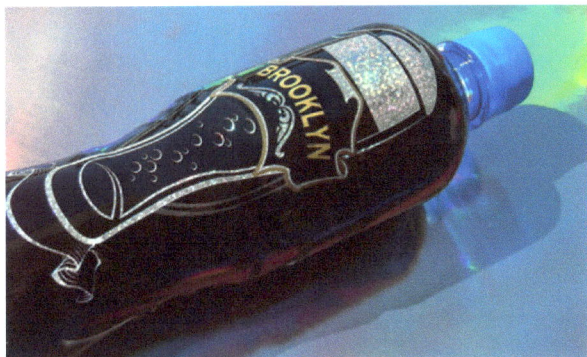

Figure 4.12 Cold foiling used on a shrink sleeve.
Source: K Laser Technology Company

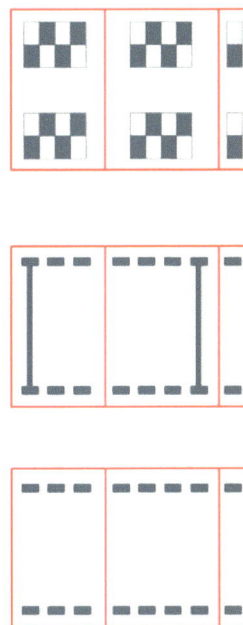

Figure 4.13 Different types of adhesive patterns

before putting sleeve products in the market. These should include having dyne solutions to **test the dyne** of the substrate, and understanding from the film supplier where the optimum dyne should be, and if corona treating is required. Once you know where the dyne level of the film should be, then treat it to the optimum level.

It is also important to do some simple **cure tests** on the printed film. Do some **tape adhesion tests**; in this industry, the standard is 610 tape, though 810 and 600 tapes are also used. Look at different tapes to make sure that the film is based on requirements, and that the tests are being passed. Some ink suppliers also advise the use of **solvent** rubs to determine cure – it is recommended to liaise with the ink supplier to determine the correct process.

Resistance tests. Bicycle crinkle tests are very important, as are block tests, or chemical resistance testing for the dairy and household markets. COF tests have already been mentioned and are also important to do. Finally, make sure to do some actual shrinking in the lab to make sure inks do not fail these same tests after being shrunk.

ACHIEVING SPECIAL EFFECTS

Apart from the conventional printing processes and techniques that have been discussed, the industry is starting to see printing being done on the outside surface of the sleeve to add a variety of special effects such as gloss, matte, and tactile varnishes that appeal to our sense of touch. Some examples include: pearlescence, iridescence, bright metallics, fluorescence, and different security features.

Some of these effects are best implemented in the lower shrink areas of the container. If **cold foiling** is to be applied (Figure 4.12) in a high shrink area, then it will probably be necessary to do a screened plate and vignette in the higher shrink areas. That way, when it does shrink, it shrinks together and it will look better. Otherwise, the foil will bunch up on itself and become dull and grey.

The same applies to **metallics**:
- Keep them out of the higher shrink areas.
- If they are going to be put in the high shrink areas, it is necessary to screen back the pruning plate and, as is done with the graphics, distort it back. This way, when it shrinks 70%, it will not pile. Good foil effect silver, which costs a lot of money, will look like a standard grey ink if it piles, which just wastes the foil and metallic effects.

Common principles concerning food packaging and migration:

The responsibility for the compliance of the packaging does not lie with only one individual member of the packaging chain.

Ultimate responsibility is with the 'person placing the pack on market' – but we all have to work together.

There shall be no unacceptable change/adulteration in the quality, odor or taste of the food/beverages.

There shall be no use of carcinogenic, mutagenic or reprotoxic substances (CMR substances).

The migration of substances evaluated for food contact shall remain below defined limits.

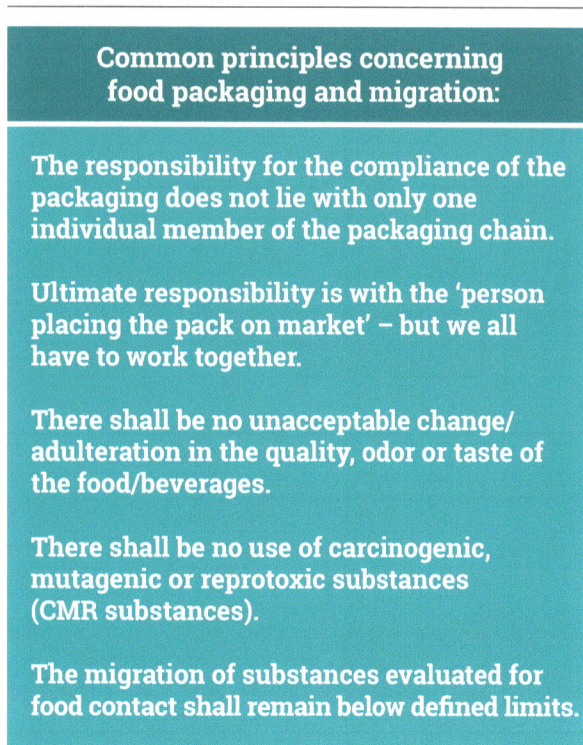

Figure 4.14 Common principles concerning food packaging and labeling. Source: Flint Group

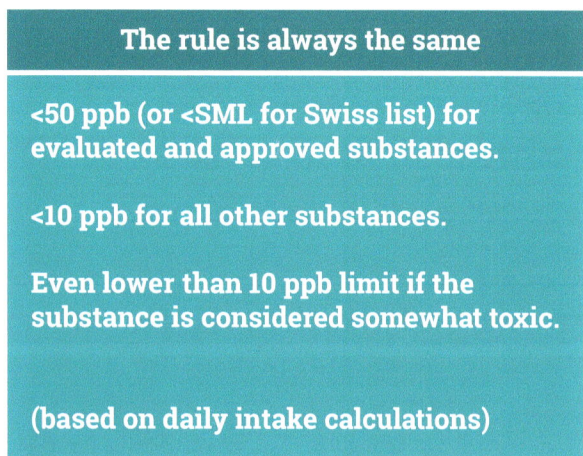

The rule is always the same

<50 ppb (or <SML for Swiss list) for evaluated and approved substances.

<10 ppb for all other substances.

Even lower than 10 ppb limit if the substance is considered somewhat toxic.

(based on daily intake calculations)

Figure 4.15 Ink migration and ink systems. Source: Flint Group

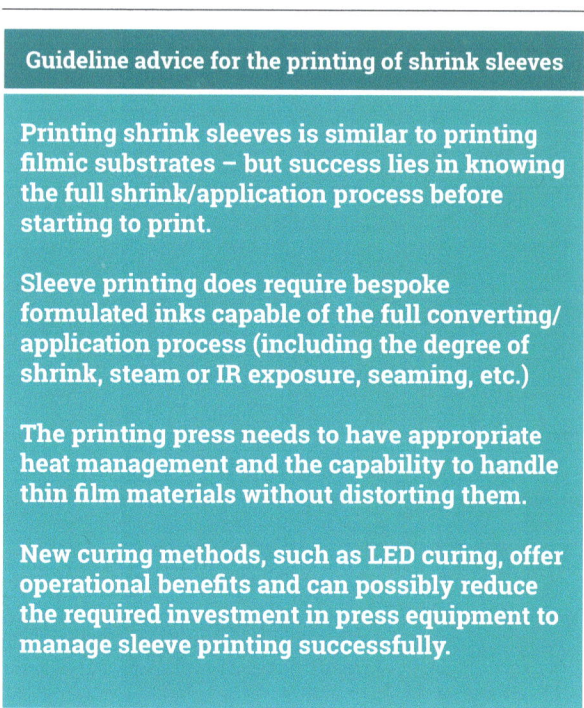

Guideline advice for the printing of shrink sleeves

Printing shrink sleeves is similar to printing filmic substrates – but success lies in knowing the full shrink/application process before starting to print.

Sleeve printing does require bespoke formulated inks capable of the full converting/application process (including the degree of shrink, steam or IR exposure, seaming, etc.)

The printing press needs to have appropriate heat management and the capability to handle thin film materials without distorting them.

New curing methods, such as LED curing, offer operational benefits and can possibly reduce the required investment in press equipment to manage sleeve printing successfully.

Figure 4.16 Guidelines for the printing of shrink sleeves. Source: Flint Group

With any of the specialty effects, make sure to talk to the suppliers as there are tricks to being successful in incorporating these types of effects. For example, if applying cold foil to a sleeve, it can be done with surface printing or it can be done with reverse printing. Reverse printing requires the use of a different cold foil than does surface printing.

Heat activated adhesives. In Chapter 2, we mentioned the phenomenon of container expansion in the tunnel and subsequent contraction after cooling, with the result being that the shrink film is loose and a cylindrical bottle will spin in the user's hand when it is being opened. To overcome this, some converters apply a heat-activated checker board or dash pattern adhesive to the film, which keeps the shrink sleeve from turning on the container, aids orientation during the shrinking process, locks the sleeve to avoid orientation while on the shelf, and even prevents the sleeve from falling off.

There are a variety of adhesive patterns that are

used, depending on the overall sleeve and container dimensions and the degree of adhesion required. See Figure 4.13.

INK MIGRATION

If sleeves are being produced for beverages or food, then it is necessary to be aware that the printed sleeve can affect the odor and taste of the product inside the container. This is not a problem with glass or metal containers, but with plastic containers it is necessary to use low migration inks.

Figure 4.14 outlines the common principles concerning food packaging and labeling, while Figure 4.15 outlines the rules governing ink migration.

It does not matter what kind of ink chemistry is used as they all have the opportunity to migrate. Ink suppliers have developed low migration chemistries that can be used for food applications, and there are many shrink applications in this segment of the market.

SUMMARY GUIDELINES

Based on the information provided in this chapter, Figure 4.16 provides summary advice to label and package printers that are considering the move into shrink sleeve printing.

A foundational understanding of ink chemistry and its components will better equip you when making ink decisions for the shrink sleeve labels you produce. Selecting the proper ink for your application, and partnering with your ink suppliers throughout the process for guidance and support, will place you in good stead to produce the perfect shrink sleeve label.

Chapter 5

___

Converting heat shrink sleeve labels – slitting, seaming, sheeting and finishing

___

With film selection, pre-press and printing steps complete, we will now pick up the process with the next step, which is slitting (Figure 5.1, center). While slitting is not new to converters, what may be new is the extent to which slitting is critical to shrink sleeve converting. What makes it more critical for shrink sleeves is that the edges of the film have a newfound importance.

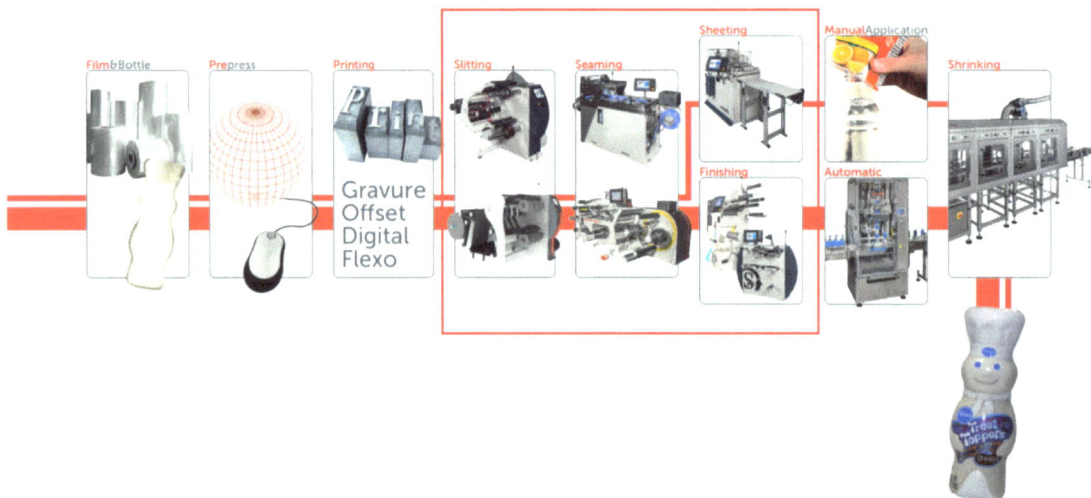

Figure 5.1 The shrink sleeve process - slitting, seaming, sheeting and finishing © 2017 Accraply, Inc

In no other slitting application are the slit edges of the rolls expected to be solvent-welded, which produces an overlapped seam that needs to be invisible on that finished product. While there is a range of slitting techniques deployed in the film converting market, several factors dictate that shrink sleeve labels be slit using tangential rotary shear technology, a concept that we will explain further in this chapter.

It is tangential rotary shear slitting that produces the cleanest cut and the least impact and stress on the slit edges of the film. This in turn maximizes our opportunity to produce a quality seam with minimal to no visibility on the finished product. Figure 5.2 represents an example of a roll with raised edges on each side of the film roll. These raised edges may be the result of one of two unsuitable slitting methods.

Figure 5.3 A crush/score slitter © 2017 Accraply, Inc

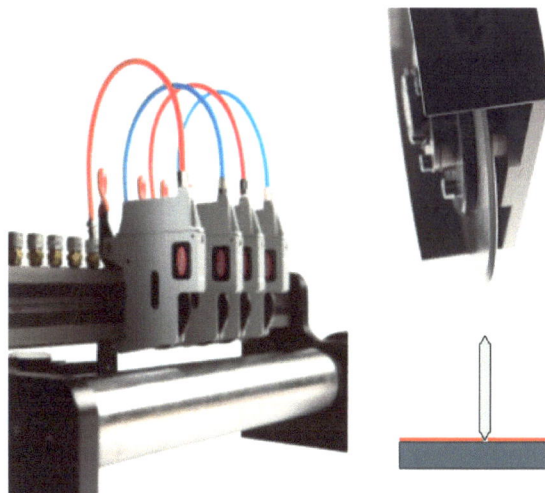

Figure 5.2 A lip is visible on either side (of the roll) © 2017 Accraply, Inc

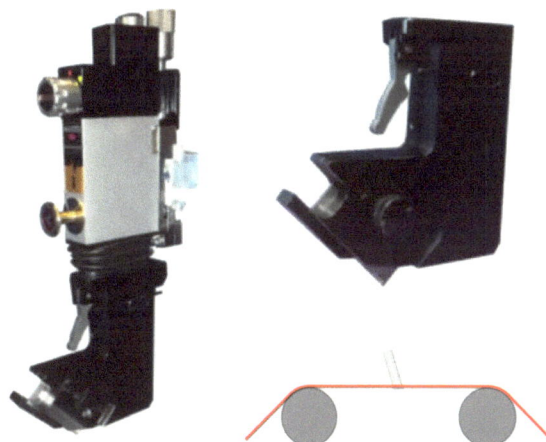

Figure 5.4 A razor-in-air slitter © 2017 Accraply, Inc

The first unsuitable slitting method for shrink sleeve film is crush slitting (or score slitting, Figure 5.3), which does not produce the clean, crisp cut to the film that the shrink sleeve process requires. The second unsuitable slitting method is razor-in-air slitting (Figure 5.4), which also produces an inferior cut to the film.

In either case, as illustrated in Figures 5.6 and 5.7, the edges of the roll have been deformed and rendered unsuitable for subsequent solvent seaming (or welding). The requirement for the finished seam on a shrink sleeve label to be as minimally visible as possible calls for flat edges and a clean cut. The only slitting method that will provide crisp, clean edges is shear slitting (Figure 5.5).

As a further illustration of the appropriateness of shear slitting, Figures 5.8 and 5.9 represent a 50x magnification of two slit edges. Figure 5.8 represents film cut using the shear slitting method, which has a demonstrably cleaner cut. Figure 5.9 represents film cut using the razor slitting method, which results in a much less crisp cut that makes the subsequent

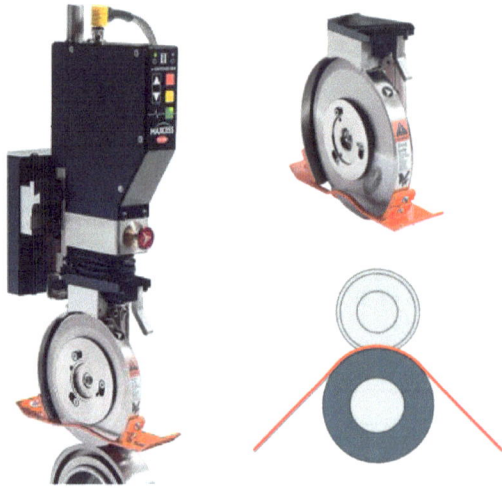

Figure 5.5 A shear slitter © 2017 Accraply, Inc

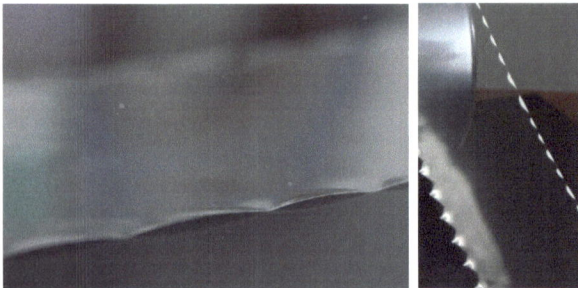

Figure 5.6 and 5.7 Examples of scalloped edges on slit shrink film © 2017 Accraply, Inc

Figure 5.8 Slit PVC film magnified 50x where shear slitting method was used © 2017 Accraply, Inc

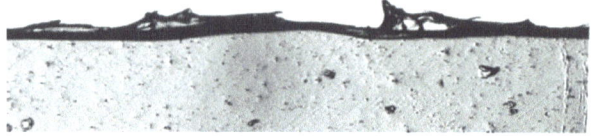

Figure 5.9 Slit PVC shrink film magnified 50x where razor slitting method was used © 2017 Accraply, Inc

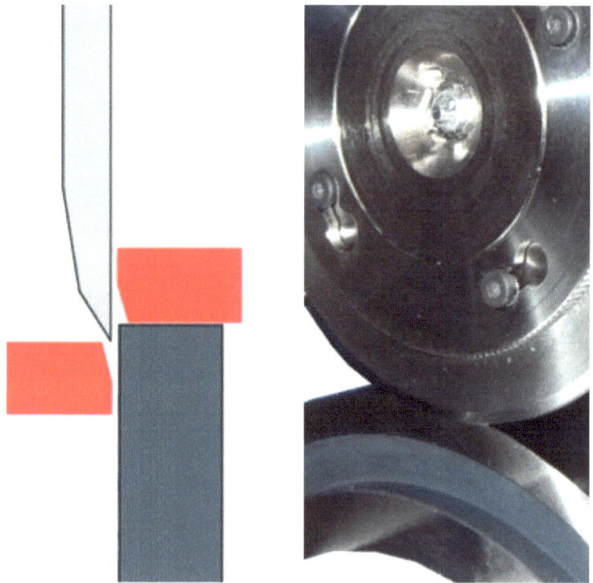

Figure 5.10 Shear slitting slices the moving web (or film) at the point where the two blades contact each other © 2017 Accraply, Inc

seaming process infinitely more difficult. Shear slitting is always the recommended method for slitting shrink film.

WHAT IS SHEAR SLITTING, AND HOW DO YOU IMPLEMENT IT?

Shear slitting is the process by which two rotating circular blades cut a moving web of film at the point where the two blades contact each other (Figure 5.10). This process is identical to the process of how cutting shears, or scissors, slice one piece of paper into two.

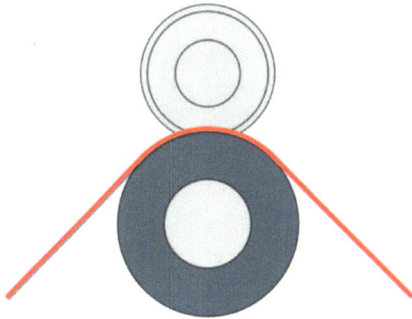

Figure 5.11 Wrap shear slitting – used for thin, flexible webs © 2017 Accraply, Inc

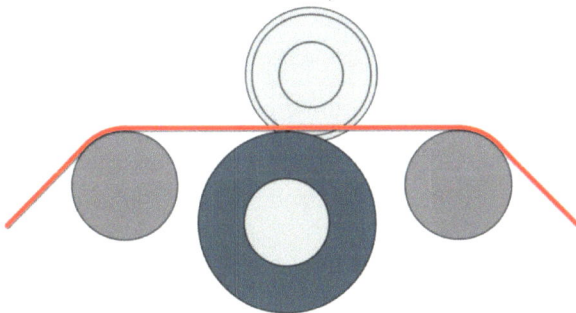

Figure 5.12 Tangential shear slitting – used for thick and more rigid forms © 2017 Accraply, Inc

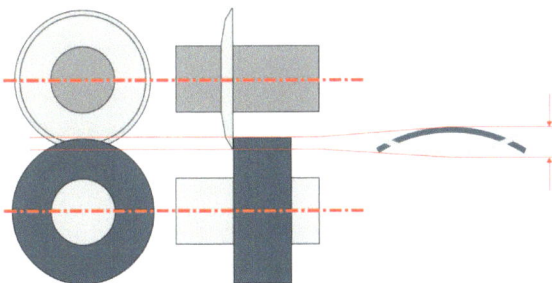

Figure 5.13 The overlap is set by how far the tangent point of the top blade is engaged beyond the tangent point of the bottom blade © 2017 Accraply, Inc

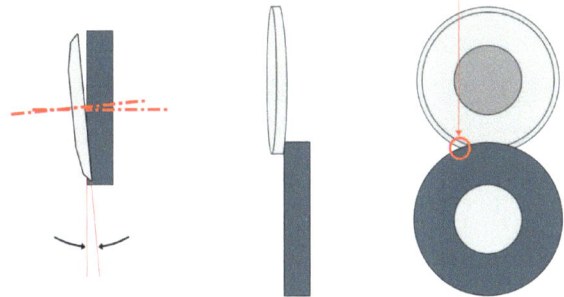

Figure 5.14 The cant angle is designed to force the shear knife contact to the overlap entrance point © 2017 Accraply, Inc

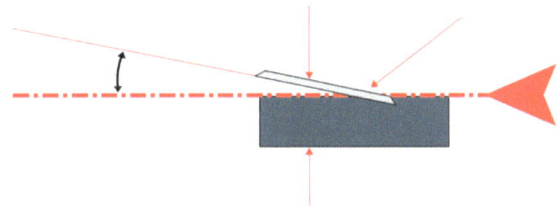

Figure 5.15 The cant angle used depends on the material being slit © 2017 Accraply, Inc

There are two types of shear slitting: wrap shear (Figure 5.11) and tangential shear (Figure 5.12). Traditionally, wrap shear is deployed for thinner materials, as the wrap curvature on the bottom blade opposes vertical deflection, preventing the web from dipping under the blade. Tangential shear slitting employs the use of entry and exit rollers, and its top blade is offset to its bottom blade. As a result, this type of shear slitting produces the same cut as wrap shear but with less risk of deformation to the edges of the film. Thus, tangential shear is preferred over wrap shear for shrink film.

Now that we have established shear slitting as most appropriate for shrink sleeve film, we now turn our attention to the five keys to shear slitting success, which are:

- **Depth (Overlap):** The depth is set by how far

Cant angle	Material
0.0 ° to 0.25 °	Metals, Plastic Sheet, Hard Web, Brittle Web
0.25 ° to 0.50 °	General Purpose Angle: Plastic Film, Laminates
0.50 ° to 0.75 °	Synthetic Fiber Products, Stretchy Films
0.75 ° to 1.0 °	Fabrics, Un-bonded, Non-woven

Figure 5.16 The predominant cant angles used for slitting different materials © 2017 Accraply, Inc

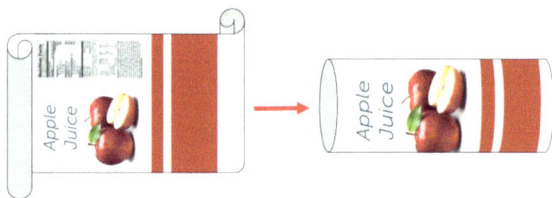

Figure 5.17 Seaming is the process of converting a flat film into a seamed tube © 2017 Accraply, Inc

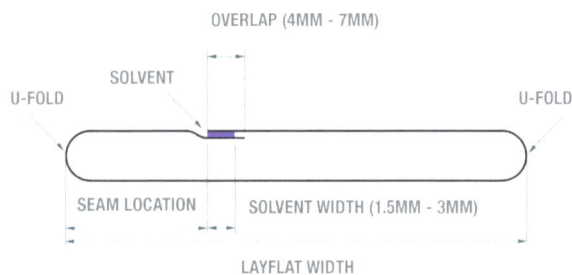

Figure 5.18 A diagram of a seamed piece of film and key terminology © 2017 Accraply, Inc

the tangent point of the top blade is engaged beyond the tangent point of the bottom blade (see Figure 5.13).

- **Cant Angle:** The cant angle forces the shear knife to contact at the overlap entrance point (see Figures 5.14 and 5.15). The cant angle used will vary depending on the type of material being shear slit, as is explained in detail in Figure 5.16.
- **Knife/Blade Profile:** The type of material being slit again determines the knife/blade profile. The primary top blade profiles used are 25, 45 and 60 degrees; 25 degrees is used for stiff, high-density webs; 60 degrees is used for thicker, lower density webs; and 45 degrees is used as the default tangential shear top blade. Satisfactory results with shrink film have been achieved using both 45 degree and 25 degree blade profiles.
- **Side pressure (force):** Side pressure, or force, is the number of pounds per square inch of pressure that the top knife places on the bottom knife. The objective here is to use the right amount of side pressure; too much can negatively impact knife sharpness and blade wear, while too little will not fully enable the bottom band to drive the top blade.
- **Knife Sharpness:** Knife sharpness is critical to create the clean slit edge that the shrink sleeve process requires. Improper cant angle positioning and the use of too much side pressure will negatively affect knife sharpness.

THE SEAMING PROCESS
Having achieved crisp, clean slits without deformation of the roll edges, we are now ready to move to the seaming (or welding) step. While our objective is to transform the flat, printed film into a seamed tube (see Figure 5.17), the goal should always be to form the perfect seam that is minimally visible and tactile. Our further goal is to do this with maximum throughput and minimal waste. Throughput is a function of speed and uptime on the equipment, while waste tends to be a function of using the right ingredients (i.e., film and solvents), equipment features and knowledgeable operators.

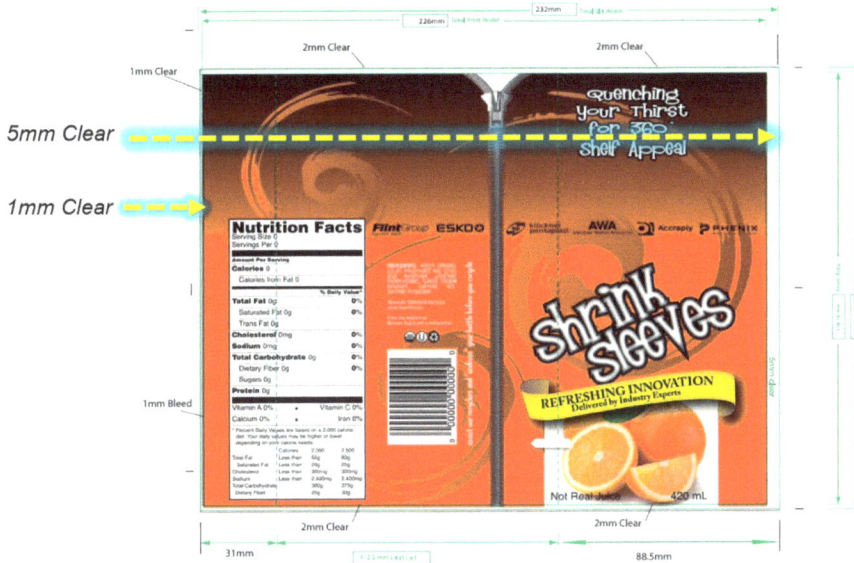

Figure 5.19 Solvent placement specifications © 2017 Accraply, Inc

Figure 5.20 A label with a vertical crack in the ink © 2017 Accraply, Inc

At this point in the process, we need to introduce some industry terminology and important concepts, which are also depicted in Figures 5.18 and 5.19:

- **Layflat:** A finished shrink sleeve that is ready for application to a container will be in the form of a flat tube. Measuring from edge to edge of this flat tube, the term used to describe this distance is called the layflat, or the layflat width.
- **Slit width:** If the layflat tube is unfolded, it provides the slit width, which describes the width of the material as it came off the slitter and entered into the seaming machine.
- **Overlap:** The overlap is the portion of the shrink sleeve tube where the seam is welded.
- **Seam location:** The layflat tube is bonded together using a solvent, which creates a seam. The position of the seam on the tube is called the seam location, and the type of container being decorated and/or the preferences of the brand owner will typically determine the seam location. Container shape is a significant determinant; the seam location on a round container may be of no importance, but with

square containers, oval containers or trigger bottles, seam location is very important. For example, it is generally preferable to avoid having a seam located on the front center panel of a product such as a household cleaning agent that is packaged in a trigger spray container.

- **Solvent:** As previously mentioned, the layflat tube is formed using a solvent weld and typically not a glue or adhesive. It is critical that the solvent and film chemistry be right to chemically bond the film material.
- **U-folds:** It is desirable to avoid crushing the edges of the material during or after the seaming process. The rationale for this is to avoid introducing fold lines that may be visible on the finished label, or worse still introduce a risk of cracking the ink, which will be visible on the finished label (see Figure 5.20).
- **Solvent placement and consistency:** To achieve the perfect seam every time, converters must be particularly mindful of the amount and placement of the solvent used to form the

55

GOOD SOLVENT PLACEMENT

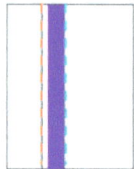

- TO THE EDGE, WITHOUT GOING OVER
- CONSISTENT WIDTH
- CONSISTENT AMOUNT

SOLVENT–PURPLE
INSIDE EDGE–ORANGE DOTTED LINE
OUTSIDE EDGE–BLUE DOTTED LINE

Figure 5.21 The solvent should be consistently applied to the outside edge, but not beyond it © 2017 Accraply, Inc

BAD SOLVENT PLACEMENT

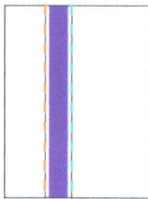

NOT TO THE OUTSIDE EDGE OF THE OVERLAP

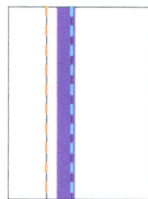

PAST THE OUTSIDE EDGE OF THE OVERLAP

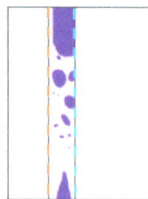

SKIPS/VOIDS

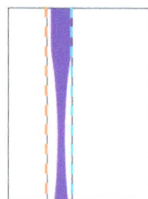

INCONSISTENT FLOW

Figure 5.22 Illustrations of the most common methods of poor solvent placement © 2017 Accraply, Inc

four commonly observed instances of poor solvent placement and consistency:

- **Not to the edge of the overlap:** When the solvent placement falls short of the edge of the overlap, the result is a coarse edge that the brand owner may reject.
- **Past the edge of the overlap:** When the solvent placement falls outside of the edge of the overlap, the result is that the solvent will bond with another adjacent layer of the film roll, which will cause 'blocking' as the roll is unwound.
- **Skips/voids:** When the solvent placement is inconsistent on the film, the seam will not fully form and will likely open or come apart on the container if not before even being placed on the container.
- **Inconsistent flow line:** Two factors can contribute to having an inconsistent flow line. The first may be due to a problem with the solvent flow control, or solvent delivery system. In the example shown, the narrowing of the solvent line may be due to an inconsistent flow of solvent due to the use of, for example, a peristaltic pump.

SEAMING STEP I HOW DOES IT WORK?

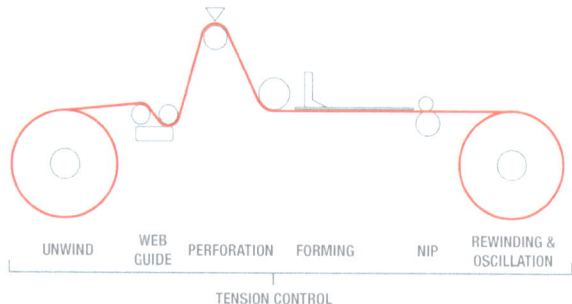

UNWIND | WEB GUIDE | PERFORATION | FORMING | NIP | REWINDING & OSCILLATION

TENSION CONTROL

Figure 5.23 The operational components of a seamer © 2017 Accraply, Inc

seam. Figure 5.21 visually demonstrates good solvent placement and consistency. More specifically, the solvent reaches to the edge of the seam without going over the edge, and a consistent amount of solvent is applied across an even width of the film. Figure 5.22 provides

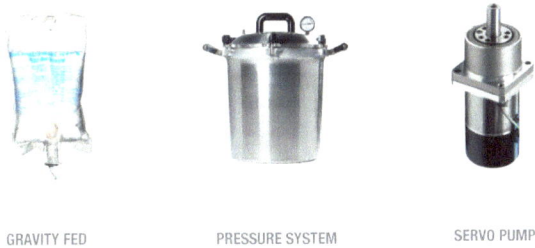

GRAVITY FED   PRESSURE SYSTEM   SERVO PUMP

Figure 5.24 Solvent control systems © 2017 Accraply, Inc

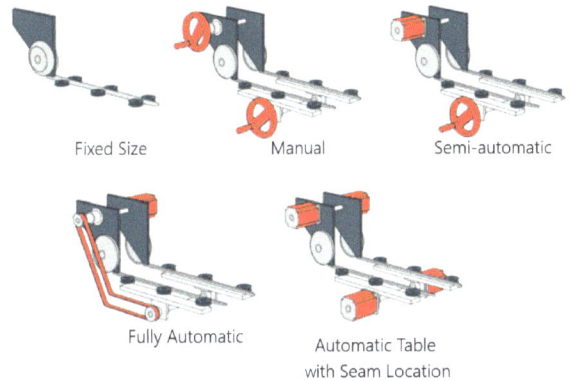

Fixed Size   Manual   Semi-automatic

Fully Automatic   Automatic Table with Seam Location

Figure 5.27 Different types of folding tooling and make-ready setups © 2017 Accraply, Inc

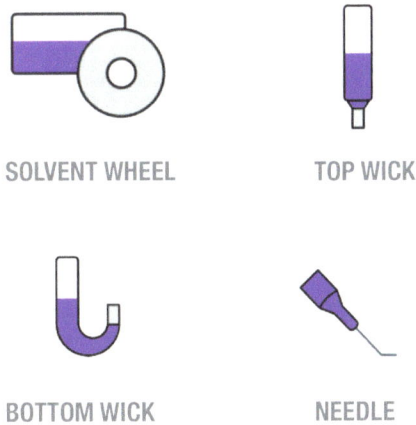

SOLVENT WHEEL   TOP WICK

BOTTOM WICK   NEEDLE

Figure 5.25 Solvent delivery methods © 2017 Accraply, Inc

THE PROCESS OF SEAMING – HOW A TYPICAL SEAMER OPERATES

The process of seaming requires a stand-alone machine that has several key characteristics, all of which are illustrated in Figure 5.23. The unwind portion of the seaming machine is where the flat material is loaded and webbed through the machine. The web guide makes corrections and adjustments to the web when necessary.

The linear perforation section, which is located just before the forming station, perforates the web in the machine direction, just before the forming process takes place. The purpose of this perforation is to give the consumer the ability to remove the sleeve from the container for recycling purposes.

The forming section is where the flat web is converted into a sleeve, or tube. This is also where the solvent is delivered to the web. The nip section is used to isolate tension, and it is also where all air is removed from the newly-formed tube before moving to the rewind section. The rewind section with oscillation is where the completed roll of seamed material is offloaded. During this process, the rewind function must have the ability to oscillate the roll. The need for this is illustrated later in Figures 5.28 and 5.29.

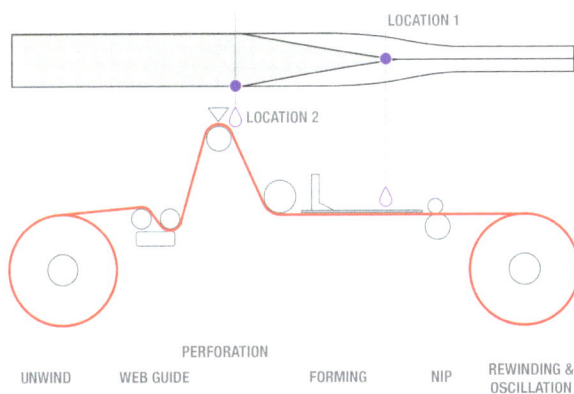

LOCATION 1

LOCATION 2

PERFORATION

UNWIND   WEB GUIDE   FORMING   NIP   REWINDING & OSCILLATION

Figure 5.26 Solvent delivery locations © 2017 Accraply, Inc

SOLVENT CONTROL

To form a shrink sleeve label, a seaming machine forms the material into a tube using a solvent, which produces a chemical reaction that welds the material together. There are three ways to control the flow of solvent to the web (Figure 5.24) when forming the material into a sleeve (or tube). They are gravity-fed, pressure system, and servo pump systems. It is important to achieve consistent results throughout the roll, and acceleration and deceleration, or running the web at different speeds, can influence the result. The amount of solvent and the rate at which it is applied, proportionate to web speed, is critical to achieve a properly formed sleeve with minimal waste.

SOLVENT DELIVERY METHODS

Solvent wheel, top wick, bottom wick, and needle are the types of delivery methods most frequently used to apply solvent to the web (see Figure 5.25). Solvent wheels are the least popular of the four. Wicks, both top and bottom, and needles are more commonly used, though both have their advantages and disadvantages. Using a wick gives the operator the ability to easily apply solvent to the edge of the seam. The problem is that wicks tend to pick up contamination and can leave striations in the seam. Some converters that print shrink sleeves using the offset printing process may actually require a wick. A fountain solution that has excessive ink contamination can cause problems for seaming solvents if ink residue is deposited in the clear seaming area. This ink residue impedes the chemical reaction between the solvent and the film, and a weak seam can result. In this case, using a wick to wipe the solvent onto the surface may help as it abrades the surface of the film and aids penetration of the solvent. High-speed seamers bring greater levels of sophistication to this process, but with the increased speed comes a need for greater solvent placement accuracy. It is for this reason that high-speed seamers typically use a needle to deliver solvent to the web.

SOLVENT DELIVERY LOCATIONS

A seaming machine can deliver solvent to the web in two locations: before the forming table while the material is still flat and stable as it passes over a roller, and on the table while the material is already formed in a tube, and just prior to it being nipped. The two solvent delivery locations are pictured in Figure 5.26. Either location is acceptable, but each has a different requirement of the seaming solvent used.

Suppliers offer a slower reacting solvent for slower machines that utilize solvent location 2. Suppliers will provide a quicker reacting solvent when using faster, higher speed machines that are delivering solvent to location 1.

Regardless of the solvent delivery location, the speed of the equipment and its ability to monitor and maintain tolerances is what will affect production quality and efficiency.

FOLDING AND FORMING

The folding and forming tooling that makes the web into a sleeve has evolved greatly over the years. Advanced folding and forming tooling design has taken much of the guesswork out of the process, as this aspect has historically been complicated and time-consuming for machine operators. Several variations of folding tooling are used in the industry (Figure 5.27). The earliest tooling used was a fixed-size shoe. Operators would insert a single, custom-sized forming shoe into the table area that corresponded to the layflat size they were running. It is rare to see this tooling used today except for high volume, dedicated production runs.

The greatest limitation to using a fixed-size shoe is that it can only be used for one specific size. Because of this limitation and the implausibility of purchasing fixed-sized tooling for every layflat, tooling evolved to accommodate the needs of converters. Manual tooling consists of forming shoes, pins and cranks, all of which are placed into position using hand levers (shown in red) that need size measuring. Semi-automatic tables utilize a hand crank, which moves the tooling into place while also notifying the operator of the exact measurements on screen. The most advanced tooling on the market is fully automatic tables. This type of sophisticated equipment enables the operator to type in the desired layflat on the screen, and the tooling will automatically move into position.

Figure 5.28 Seams fall in same place © 2017 Accraply, Inc

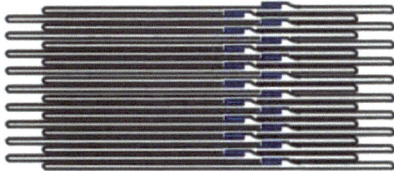

Figure 5.29 Rewind oscillation distibutes the seam on the roll © 2017 Accraply, Inc

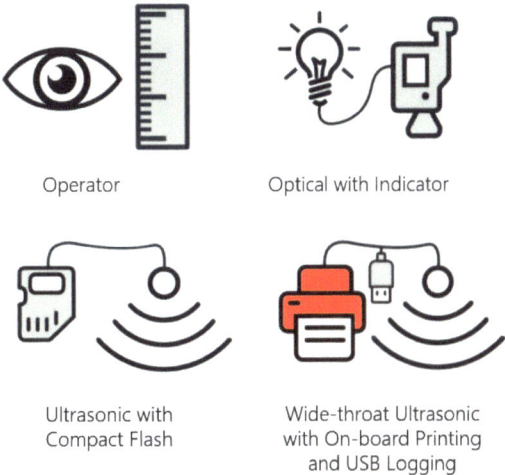

Operator

Optical with Indicator

Ultrasonic with Compact Flash

Wide-throat Ultrasonic with On-board Printing and USB Logging

Figure 5.30 Various types of monitoring and control systems © 2017 Accraply, Inc

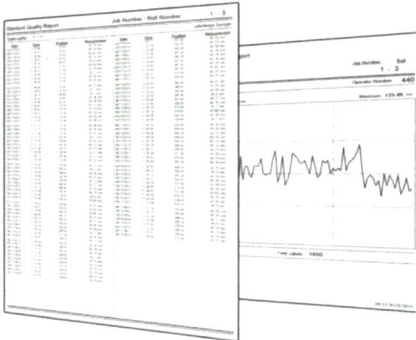

Figure 5.31 Certificates of analysis © 2017 Accraply, Inc

REWIND AND OSCILLATION

An oscillated rewind is an important component of the shrink sleeve process. When examining a shrink sleeve, one will observe that the thickness is tripled at the seam reaction area. If this finished roll of shrink sleeve labels is rewound in the traditional sense, all of the tension would build up in that seam area (Figure 5.28). The reel is then likely to telescope and topple over, wasting the production time and material used in this process. The seam therefore needs to be spread back and forth, which is referred to as oscillation (Figure 5.29).

Some co-packers and end users require a specific amount of oscillation, but a good rule of thumb for operators and customers to follow is to oscillate twice the overlap.

MONITORING AND CONTROL

Historically, operators would start and stop their machine on a regular basis to collect, assess and measure product samples. The industry then moved into optical monitoring with indicators, which still required the operator to assess and monitor performance while the machine was running at production speed. Today, the industry relies on ultrasonic sensing technology; an ultrasonic sensor, with proper software, can identify layflat webs running through the equipment, monitoring and qualifying the layflat size (Figure 5.30). A more recent advance gives the operator the ability to download the production data to a USB drive as well as print said data for later reference. Furthermore, many decorators require a certificate of analysis (Figure 5.31) to show the

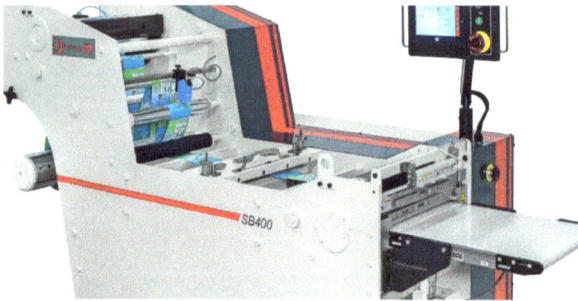

Figure 5.32 Perforation must be in register for the sheeting process © 2017 Accraply, Inc

SHEETING

Following the finishing steps of the seaming process, the labels will either be applied to the container automatically or by hand. If the labels will be applied automatically with label application equipment, then the finished product will remain in roll form. If the labels will be applied by hand, then the labels will need to be cut into individual sleeves. A sheeting machine is required to produce individual cut labels. Should manually applied sleeves require perforation, then that too must be inserted using this equipment (Figure 5.32).

PERFORATIONS IN SHRINK SLEEVES

The primary purpose of shrink sleeve perforation is for tamper evident sealing, and this perforation is typically performed on a seaming machine with linear perforation as previously outlined in Figure 5.23. Examples of shrink sleeve perforations for tamper-evidence can be seen in Figure 5.34.

The secondary purpose of shrink sleeve perforation is for recyclability, or more specifically, to provide the consumer with an easy means of removing the shrink sleeve label from the container so that the container can be recycled.

Tamper evident labeling (Figure 5.35) is also used for tax strips and anti-counterfeit purposes. These labels typically incorporate linear perforation produced on the seam machine, and they are often combined with a tear-strip or holographic material, which is

decorator that the product was monitored and qualified.

THE FINISHING STEPS

The finishing steps of the seaming process consist of unwinding and rewinding the seamed web using a Doctor Machine® with oscillation, changing the core size (when necessary), changing the application rewind direction (when necessary), and building finished sized rolls. It is also at this step that splices in the seamed rolls are 'repaired' and made suitable for passing through the automatic label application equipment, before being detected and ejected from the line prior to the shrink tunnel.

Linear Perforation   Sheeter- Perforation   Stand-Alone Unit   Integral Unit

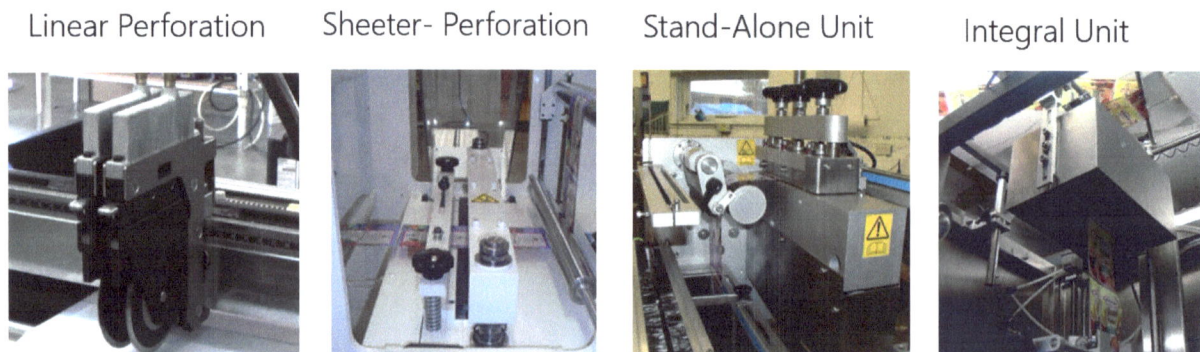

Figure 5.33 The different types of perforation units © 2017 Accraply, Inc

Figure 5.34 Examples of shrink sleeve perforation for tamper-evident sealing © 2017 Accraply, Inc

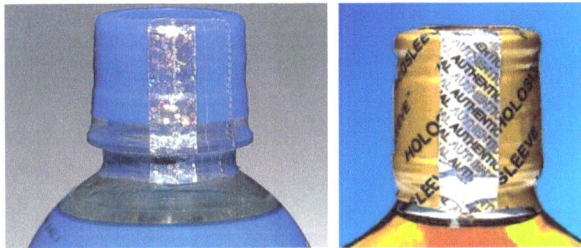

Figure 5.35 Tamper evident labels incorporating tax strips and anti-counterfeit holographic materials © 2017 Accraply, Inc

generally an easy add-on to seaming equipment (Figure 5.36).

There are also other types of tamper evident sleeves with tear tabs that come down the container

Figure 5.36 Tear strip application device

wall and then perforate around the container, and are typically used for a full body sleeve that also requires a tamper evident seal. Sometimes a brand owner will require a partial body tamper evident seal where the T-perforation is used around the container lid while still maintaining the container's brand identity on the neck and body of the container.

Most of this is done on the automatic application equipment unless it is going to be sheeted for hand application, in which case, the perforation is completed on the sheeting equipment using a perforation device.

CONCLUSION

The secondary converting steps materially affect the quality of the finished label on the shelf, and each step requires great attention to detail. It is therefore of critical importance to use the best ingredients – film, ink, solvent, equipment, and trained people – to ensure the production of the highest quality, most consistent shrink sleeve label in the market.

Chapter 6

———

Shrink sleeving – using the right application and shrink tunnel technology

———

We have spent most of this book reviewing the process of shrink sleeve manufacturing, beginning with materials selection, then moving through origination, ink and printing, and the secondary converting steps. Now we will take a closer look at the process of sleeve application and how the sleeve is shrunk onto the container (see Figure 6.1).

———

While converters will be most interested in how to produce a high value shrink sleeve label, it is also necessary to understand that the sleeve application process is as important as the previously mentioned steps in order to produce a finished product that the brand owner – and ultimately, the market – will value. Converters should always be tightly connected with the manufacturer of the application equipment in order to understand the specifications and tolerances required of the specific application equipment being used by each of their customers. In other words, there are no one-size-fits-all solutions in the successful application and shrinking of shrink sleeve labels.

Prior to reviewing the shrink sleeve application machinery in some depth, let us review some of the key concepts that will bear importance in the application process.

The layflat size and its tolerance – as well as the extent to which the layflat is over-sized relative to the

container – is important because it determines the compatibility of both the container and the tooling on the application machinery with the shrink sleeve. The oversize is typically 2mm on a round container, but it may be as much as 7mm, depending on the container shape, size and the method of application.

The repeat length, or cut length, describes the length of the label being applied to the container. Each sleeve has a clear, non-printed area on the top and bottom. These clear areas are used for both manual and automatic application purposes. In manual application, the clear areas act as a guide for an automatic sensor to determine where to cut, or sheet, the individual sleeves. In automatic application, the clear areas also indicate where to cut the sleeve just prior to applying it onto the container. It is therefore always necessary to have these clear areas on any finished shrink sleeve.

Let's now return to the concepts of smiling and frowning (Figure 6.2) that were first discussed in

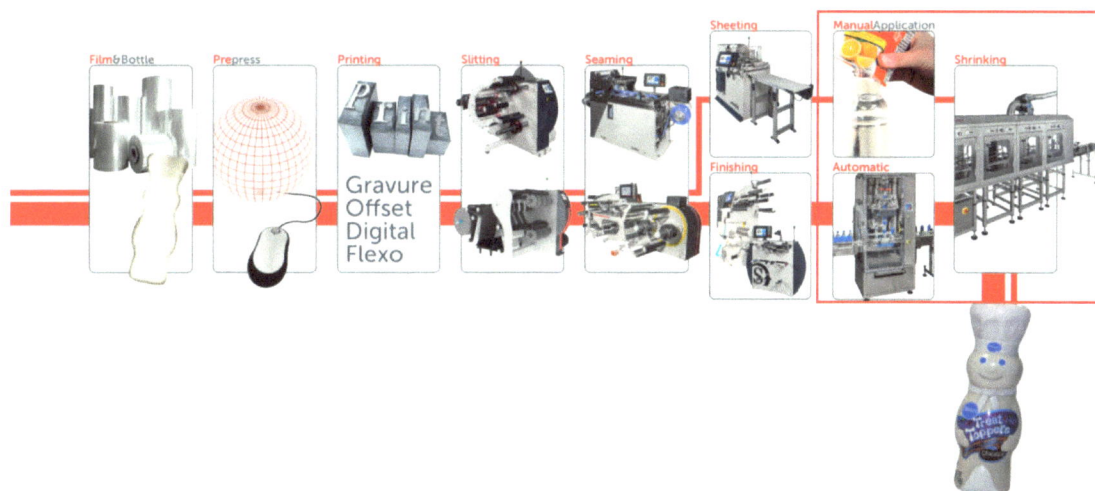

Figure 6.1 The shrink sleeve process - slitting, seaming, sheeting and finishing © 2017 Accraply, Inc

Chapter 2. Smiling and frowning comes about when the sleeve shrinks in both the machine direction (MD) and the transverse direction (TD), and an arching effect is created.

Fortunately, there are ways to overcome this issue. One method is to carefully select the type of film used and control for how the film shrinks into position on the container. Another method to control the issue of smiling and frowning labels is to incorporate machinery that lifts the container off the conveyor and allows the sleeve to shrink under the base of the container. Finally, a third method of mitigating smiling and frowning labels is through container design; specifically, either selecting or designing a container with a built-in notch to lock the sleeve label into place as it shrinks.

Figure 6.2 An example of shrink sleeve frowning © 2017 Accraply, Inc

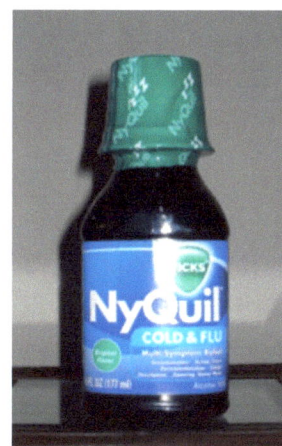

Figure 6.3 A tamper evident band © 2017 Accraply, Inc

SHRINK SLEEVE LABEL TYPES

There are several types of shrink sleeves that brand owners seek for their products. Tamper evident or tamper bands (Figure 6.3) are used to produce a seal on a container that demonstrates tampering if the container is opened.

A part-body sleeve (Figure 6.4) is used when a container does not require a full-body decoration. A brand owner may require us to position a sleeve half

way up a container or at the top of a container.

One of the challenges with positioning a part body sleeve half way up a container is that it becomes necessary to somehow anchor the sleeve in the right position on the container. Some of the containers in Figure 6.4 incorporate a notch, which allows the clear area to shrink into the groove and provide a nice finish.

Figure 6.4 Examples of part-body sleeves © 2017 Accraply, Inc

Figure 6.5 A full-body sleeve with perforation © 2017 Accraply, Inc

Full body sleeves, on the other hand, allow for 360-degree and top-to-toe container decoration. No other labeling technology allows for almost 100% of the container's surface to be decorated and used to convey the brand's message to the consumer.

It is also possible to have full body decoration but with the added functionality of a perforation (Figure 6.5), which aids the consumer in removing the label following consumption. This feature can be achieved in the label application process, as well as during the converting process (which precedes label application). It is particularly straightforward to achieve on the application equipment, as the tooling involved is easily installed and generally provides a lot of flexibility in the positioning and pattern of perforation.

SHRINK SLEEVE LABELING PERFORMANCE CONSIDERATIONS

Achieving success in the final steps of applying and shrinking a shrink sleeve label hinges on decisions made at the very beginning of the process – decisions like what material the container will be made from and what film material was chosen for the sleeve. For example, glass containers present a very different scenario at the shrinking stage than do containers made from HDPE. Process conditions are important too – there are fundamental differences between shrinking onto a container that is full versus one that is empty.

Figure 6.6 captures some of the main variables that are in play as we set about applying and shrinking a finished shrink sleeve. Without the right combination of container type, film type and process conditions, a successful result will prove difficult to achieve. This, of course, assumes that all the other steps in the process have been completed with the appropriate focus on quality.

Container material considerations: Since glass is a heatsink, it can present challenges in the shrink tunnel by effectively stopping the film from shrinking evenly when the warm film comes in to contact with the colder surface of the glass. It is for this reason that glass containers are frequently pre-heated prior to applying the sleeve. PET containers, particularly if empty and lightweight, can be a challenge with regards to stability for sleeve application, and are susceptible to shrinking in the tunnel, along with the sleeve. On the other hand, HDPE containers expand in the tunnel – particularly when the containers are empty – and result in the sleeve shrinking around an expanded container, that then becomes loose-fitting after the HDPE contracts upon cooling. Metal containers behave similarly to glass in that they are a heatsink, while paperboards (as used for ice cream, for example) may require chilled conveyors and careful targeting of the heat in the tunnels to shrink the tamper band without melting the contents of the container!

Process conditions: The most fundamental starting point is with whether the container is filled or empty. Filled containers may be more stable on the line, but they may also have over-spill from the filling process that may require washing prior to sleeving. On the other hand, empty containers do not run the risk of over-fill contamination, but they do present greater instability on the line and have higher risks associated with their shrinkage or expansion in the tunnel. Filling temperatures can also present challenges. While ambient filling temperatures are generally straightforward, cold-filled products bring issues of condensation on the surface of containers,

BEFORE STARTING | PERFORMANCE CONSIDERATIONS

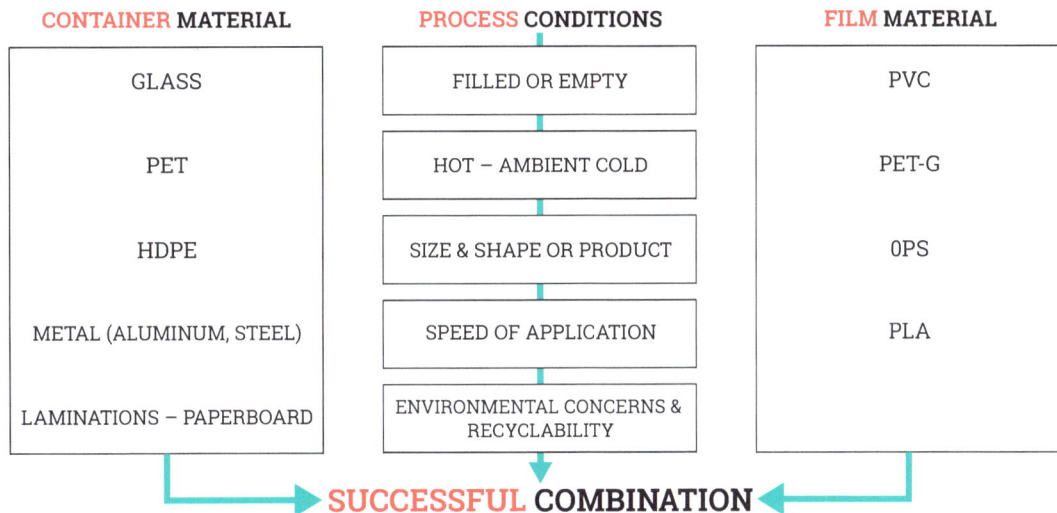

CONTAINER MATERIAL	PROCESS CONDITIONS	FILM MATERIAL
GLASS	FILLED OR EMPTY	PVC
PET	HOT – AMBIENT COLD	PET-G
HDPE	SIZE & SHAPE OR PRODUCT	OPS
METAL (ALUMINUM, STEEL)	SPEED OF APPLICATION	PLA
LAMINATIONS – PAPERBOARD	ENVIRONMENTAL CONCERNS & RECYCLABILITY	

SUCCESSFUL COMBINATION

Figure 6.6 Key performance considerations associated with shrink sleeve labeling © 2017 Accraply, Inc

Figure 6.7 A trigger container

making it difficult to have sleeves slide down the container. Conversely, hot-filled products may cause the sleeve to pre-shrink on the container during application.

All the foregoing comments also have bearing on the **film selection** – with the advantages and disadvantages of each film type covered in Chapter 2. Each of the three components is important to understand, because if there are issues with any one of the three elements, an inferior shrink sleeve label may result.

APPLICATION TECHNOLOGIES

Having understood that there are multiple container types, process conditions and film material combinations called for in the marketplace – and that ultimate success depends on selecting a combination of these three elements that are compatible with each other, let's now review various options for applying the shrink sleeve label to the container. Essentially, there are three heat shrink sleeve application technologies available to us:

1. Rotational (carousel) systems
2. Direct apply systems
3. Mandrel systems (sometimes called bullet-style systems).

Carousel systems offer the advantage of facilitating a 'squaring' of the sleeves, which allows for better label placement, or rotational accuracy, on non-round containers. A carousel system will often be used to apply shrink sleeve labels to containers such as trigger bottles, like that pictured in Figure 6.7. The formation of square or rectangular sleeves prior to their application to non-round containers greatly

Carousel/Rotary Systems

Simplified change parts

Low to intermediate speed

Non-round/shaped containers

More tolerant of lower quality sleeve materials

Better suited to frequent changeovers

Figure 6.8 A carousel (or rotary) shrink sleeve application system and its benefits © 2017 Accraply, Inc

Figure 6.9 A carousel sleeve applicator in operation © 2017 Accraply, Inc

Direct Apply Systems

Low to intermediate speeds

Low technology change parts

Compact design with a small footprint

Figure 6.10 Direct apply systems © 2017 Accraply, Inc

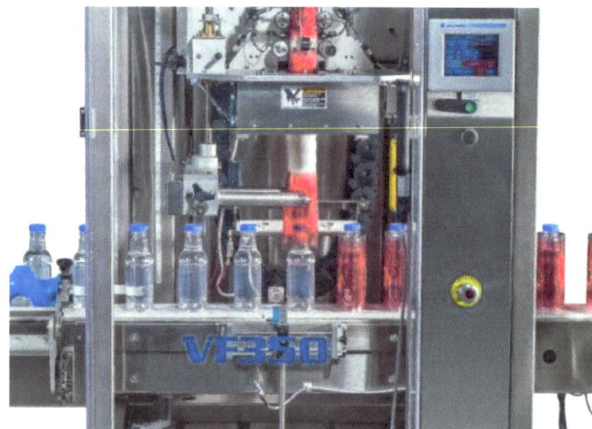

Figure 6.11 A vertical mandrel sleeve applicator in operation © 2017 Accraply, Inc

increases the probability of the label panels being correctly positioned on the container after shrinking.

You will notice from Figure 6.8 that carousel application systems are more tolerant of lower quality sleeve materials. This tolerance can be attributed to the fact that the carousel tooling comes into contact with the sleeve significantly less than is the case with a mandrel or 'bullet' type application system, thus the tolerance and fit are less of a concern with a carousel application system.

Figure 6.9 shows a carousel applicator machine in operation. The carousel system shapes the sleeve in such a fashion that it is well-positioned or centered on the non-round container as it is applied.

Direct apply systems are used primarily for tamper band applications. The process is simple: the sleeve passes through the tooling, where it is opened up into a round shape. As the container being sleeved moves along the conveyor, a guillotine cuts the sleeve, and it is dropped onto the container as it passes.

Vertical/Mandrel Systems

Change parts are more sophisticated

Intermediate to high speed

Suited to round or cylindrical containers

Quality materials required

Often chosen for continuous 24/7 operation

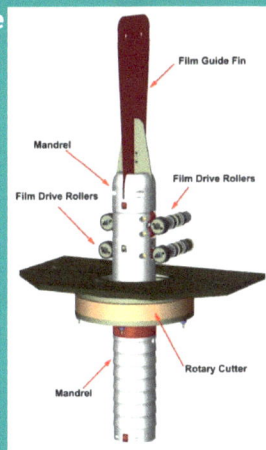

Figure 6.12 The vertical mandrel sleeve application system © 2017 Accraply, Inc

Hot Air Shrink Tunnels

Advantages:
- **Versatile**
- **Cost effective**
- **Directional heat application**
- **Good for focused heat on necks, recesses & grooves**

Disadvantages:
- **Air does not transfer heat very efficiently**
- **Filled/cold products can act as a heatsink and impede shrinking**
- **'Heat shadows' can cause unhelpful distortion and uneven shrinking**

Figure 6.13 Advantages and disadvantages of hot air shrink tunnels © 2017 Accraply, Inc

Direct apply systems (Figure 6.10) are very simple to use, they generally use low-tech change parts, and from an operator's point of view, they offer simple and quick changeovers.

Mandrel (or bullet-style) application systems are the most widely used style of application systems in the market. This system uses a mandrel, or cylinder-like tube, around which the sleeve is opened (see Figures 6.11 and 6.12). The mandrel is suspended within a system of drive rollers that move the opened sleeve down the mandrel before the sleeve is rotary-cut to fall (or be driven) directly onto the moving container. With this type of application, the mandrel comes into contact more extensively with the sleeve than does the tooling on a carousel application system. The layflat size tolerance – as well as the coefficient of friction (COF) characteristics of the inside of the sleeve – are much more critical on the mandrel style system than is the case with carousel or direct-apply systems.

With mandrel application systems, the speed at which the container is moving must be accurately timed with the mandrel as it fires the sleeve off and onto the container. It is for this reason that we use mandrel systems primarily for round or cylindrical containers as they present as better 'targets' for the sleeve as it is leaves the bottom of the mandrel, usually at high speed.

SHRINK TUNNELS

There are three main types of shrink sleeve tunnels:

1. Hot air
2. Radiant heat
3. Steam

This section will summarize the technologies available and discuss the advantages and disadvantages of each.

Hot air tunnels (see Figure 6.13) are versatile, they are cost-effective to use, and they can connect to almost any power source. Hot air tunnels offer directional heat, so depending on the type of equipment used, the many different manifolds in these systems enable heat to be focused on those areas of the container that require the most shrinkage. This makes them good systems for focused heating on necks, recesses and grooves.

Radiant Heat Tunnels

Advantages:
• Versatile
• Very high energy input possible
• Excellent for pre-heating containers

Disadvantages:
• Lacks directional discretion
• 'Heat shadows' can influence results
• Sleeve artwork and color variations can influence results
• Requires careful use with certain containers

Figure 6.14 Advantages and disadvantages of radiant heat tunnels © 2017 Accraply, Inc

Steam Tunnels

Advantages:
• Versatile
• Optimum shrink quality performance
• Gives the greatest process window

Disadvantages:
• Additional capital outlay for steam source and associated equipment if not existent - piping, extraction, valve-gear, water, drainage
• Consider risk of residual internal/external moisture
• May require addition of heat guns on some containers
• May require addition of air knife to dry/de-water containers

Figure 6.15 Advantages and disadvantages of steam shrink tunnels © 2017 Accraply, Inc

Hot air tunnels do have some disadvantages, however. Air is not a very efficient medium through which to transfer heat. Therefore, temperatures in hot air tunnels are generally higher to ensure enough heat is transferred to the film to start the shrink process. Because of the higher temperatures, the leading edge of the container – or more specifically, the sleeve on the container – can be over-exposed to heat as it enters the tunnel, resulting in an uneven shrink finish. This can be particularly prevalent with certain container types and process conditions, such as with cold-filled plastic or glass containers. One way to mitigate this issue is to rotate the container as it passes through the tunnel using a spinning conveyor.

Radiant heat tunnels (see Figure 6.14) were primarily designed for preheating glass containers – prior to sleeving – to mitigate the heatsink effect of the glass. However, they can also be very effectively used to shrink sleeves. Radiant systems deliver infrared heat and because the heat remains in the chamber, they create an oven-like shrinking environment.

Due to their high temperature of operation, radiant heat tunnels represent a very harsh shrinking environment with little opportunity to direct heat toward specific areas on a container. Additionally, it can be particularly difficult to get even, consistent shrink results. For instance, when the leading side of the sleeve entering the tunnel gets aggressively shrunk before the trailing edge, or when the sleeve on the two sides of the container are exposed to more intense heat than the leading and trailing sides, a 'pulled' or uneven finished shrink sleeve may result. A further complication of the high temperatures in radiant tunnels is the difficulty they can present with empty containers. For example, the task of shrinking a PET sleeve on a thin-walled empty PET container can be extremely challenging in a radiant tunnel.

Steam tunnels (see Figure 6.15) are the preferred medium for most shrink sleeve applications, and they offer some significant advantages in terms of the process window. Steam tunnels are the most versatile type of tunnel and work well with a variety of films. Steam distributes heat very evenly to the entire surface of the film as it envelopes the container passing through the tunnel. And, because it is steam – and water is over 20 times more efficient at transferring heat than is air – temperatures are lower and the environment in the tunnel is much less harsh.

Despite steam tunnels being the most favored method of shrinking shrink sleeve labels due to their propensity to deliver the most even finished result, their initial installation is likely to be more involved and expensive due to the need for a steam generating boiler, as well as all the associated piping, extraction, valve gear and drainage required. The volume of steam – and hence the boiler requirements – will primarily be dictated by the volume of throughput required of the sleeving line.

It is common for containers to emerge wet from a steam tunnel. In situations where this presents a problem (e.g., with secondary packaging), air knives may be required to dry the containers.

In summary, this chapter has focused on the final steps in the shrink sleeve production process – applying the sleeve to a container, and shrinking it. As with all previous steps in the process, a focus on detail and a determination to make informed decisions and produce quality at each step throughout the process is essential to achieving perfection on the retail shelf. There are no one-size-fits-all solutions in the production of shrink sleeves as each container and each label produces its own set of challenges. Every decision that is made – right from the point of selecting the container shape and what material it will be made from – has influence on the best method of applying and shrinking that label at the end of the process. And, the earlier you involve your application and shrinking equipment manufacturer in the decision-making process, the greater your likelihood of success.

Chapter 7

Challenges, learnings and the quest for perfection

The preceding chapters of this book have been organized around the fundamentals of the production process for shrink sleeve labels. The Foreword challenged us to understand how to make the *perfect* shrink sleeve label, and the subsequent chapters provided us with a depth of knowledge around each step of the production process. This chapter serves to use real-world examples of projects that have fallen short of the perfection we seek, categorized into six key areas – container selection and shape, film selection, graphics, ink selection, slitting and seaming, and sleeve application and shrinking. By using the previous chapters to provide you with the requisite knowledge, and setting up this chapter to illustrate some real-world learnings, it is our hope to drive you incrementally closer to achieving perfection in the shrink sleeve production process.

CONTAINER SELECTION AND SHAPE

Shape plays an important role in the shelf appeal of a shrink sleeve label. The most impactful sleeved products tend to be those that effectively – and simply – combine shape with graphics.

The Pillsbury Doughboy in Figure 7.1 is a great example of a very effective combination of shape with graphics.

The teddy bear in Figure 7.2 is equally impactful, and emphasizes the value of shape. The brand owner invested significantly in a custom glass container, incorporating all the physical features of the bear.

Shape does not always need to be exotic to be very effective, however. The high-gloss, extremely well-executed Sunsweet® prune container in Figure 7.3 is incredibly eye-catching. It oozes quality and is entirely practical, incorporating tamper evidence into its design.

Shape can also work against us, as we can observe in the next two examples. The herb grinder in Figure 7.4 is dressed in a self-adhesive label with a tamper evident sleeve. Because the container shape is devoid of a lip or a groove under or into which the tamper evident sleeve could be shrunk, the tamper evident sleeve can in fact be removed – and replaced – without any evidence of tampering!

As versatile as the shrink sleeve labeling technology is, it is not a given that it can or should be applied to every container shape. The Propel bottle in Figure 7.5 is effectively dressed in a shoulder shrink

Figure 7.1 The Pillsbury Doughboy – an exceptional example of shape and graphics © 2017 Accraply, Inc

Figure 7.2 A glass container shaped to match the graphics © 2017 Accraply, Inc

Figure 7.3 Simple shape and vibrant graphics create impact © 2017 Accraply, Inc

Figure 7.4 Shape rendering the tamper evident sleeve ineffective © 2017 Accraply, Inc

sleeve label – and the horizontal groove around the bottle has provided an effective means of anchoring the sleeve at the desired location. Conversely, the entirely smooth and gentle curves of the H2Oh! bottle provide no such anchoring – resulting in an inconsistent and unattractively decorated bottle on which the sleeve is crooked.

The material from which the container is made also has impact on the quality, look and feel of the final product on the retail shelf. Glass, for example, is a hard, rigid material (compared to plastic), and the sleeves on glass containers tend to get damaged as the containers bump against each other on conveyers or during transport. Damage such as

Figure 7.5 Grooves or ridges provide excellent anchor points for partial-body shrink sleeves © 2017 Accraply, Inc

Figure 7.7 An HDPE container that expanded and subsequently contracted, resulting in a loose-fitting sleeve © 2017 Accraply, Inc

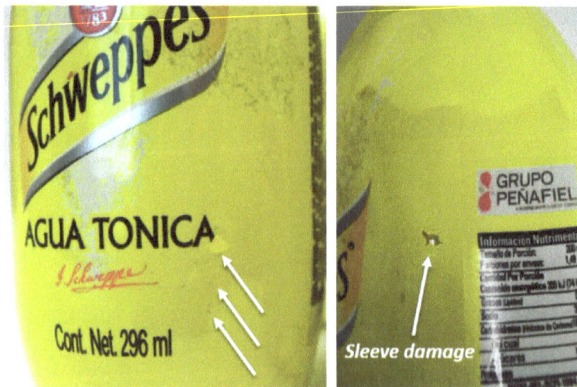

Figure 7.6 Glass bottles with damaged sleeves due to rubbing against each other in transport © 2017 Accraply, Inc

heat-activated adhesive to the inside of the sleeve in order to prevent the sleeve from rotating on the container, due to its looseness.

THE IMPORTANCE OF FILM SELECTION

We covered the importance of film selection in Chapter 2, and you will note that this has been a recurring theme throughout this book.

Figures 7.8 and 7.9 illustrate unfinished shrink in the high shrink areas of the container. While this may have been due to issues in the shrink tunnel, it bears all the hallmarks of an incorrect film selection. In other words, the film selected did not have enough shrink capability to form a perfect fit in the high shrink areas.

It is important to realize that shrink film is produced with a specific maximum shrink percentage, with a tolerance that varies by film supplier. Applying extra heat will not cause the film to shrink beyond that which it has been engineered to do. Furthermore, when selecting a shrink film for a particular container, it is important to incorporate a safety margin consistent with the shrink percentage tolerance offered by the film supplier – typically 2% to 5%. Allowance should also be made for the fact that printed film will not shrink to the same extent as raw film.

that evident in Figure 7.6 is frequently observed with glass containers.

While PET containers are among the most straightforward and flexible to sleeve, HDPE can be quite complex. As mentioned in Chapter 6, HDPE expands in shrink tunnels and contracts upon cooling. So, sleeves that shrink to the expanded HDPE in the tunnel become loose-fitting once the container cools. This can result in a less-than-ideal presentation of the finished product, as is demonstrated in Figure 7.7. On round HDPE containers, it is quite common to add a

Figure 7.8 A film with insufficient shrink potential to finish over the cap © 2017 Accraply, Inc

Figure 7.9 A film with insufficient shrink potential to finish over the neck © 2017 Accraply, Inc

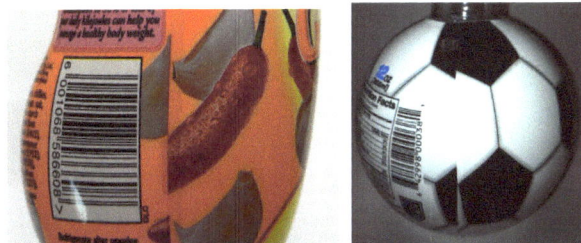

Figure 7.10 and 7.11 Sleeved containers where the designer neglected to consider the graphics coming together at the seam © 2017 Accraply, Inc

THE IMPORTANCE OF GRAPHICS

Graphic design is clearly a major element in the appeal and success of the shrink sleeve labeling format. Graphic appeal – and the successful marriage of graphics with shape – is simply foundational.

The most fundamental concept often missed in the design of shrink sleeve labels is the realization that the label is three-dimensional. It is not a flat label, like a self-adhesive or a roll-fed label.

Figures 7.10 and 7.11 are great examples of that which happens when a designer approaches shrink sleeve design with a two-dimensional mentality. They forget that their design will come together, or overlap, at the seam. With just a little forethought, the graphic of the chilies and garlic in Figure 7.10 might not have been cut off as they have been – and furthermore, the designer could have avoided drawing so much attention to the seam, which is otherwise fairly inconspicuous.

The same is true of the soccer ball bottle in Figure 7.11. It would have cost nothing extra to have had a more complete and attractive finish at the seam. In contrast, the graphics in the seam area of Figures 7.12, 7.13 and 7.14 have been designed with a three-dimensional mindset. The seams are not accentuated, but rather camouflaged by the designer.

The location and orientation of the barcode is another seemingly small detail that is frequently missed in the design of shrink sleeve labels. The first rule of thumb calls for barcodes to be placed in low shrink areas. The second rule of thumb suggests that they be vertically configured using ladder-style orientation. This orientation runs a lesser risk of distortion, as most shrink occurs in the transverse (i.e., horizontal) direction. Figure 7.15 illustrates the horizontal placement of a barcode that is likely unreadable due to the extent of the transverse direction shrink. Figure 7.16 illustrates the correct barcode placement and orientation.

A final point on graphics – make use of the entire canvas! Figure 7.17 illustrates a shrink sleeve label being used to replace a front and back self-adhesive label – that looks exactly like a self-adhesive label! The full opportunity presented by the shrink sleeve format has been entirely missed in this case.

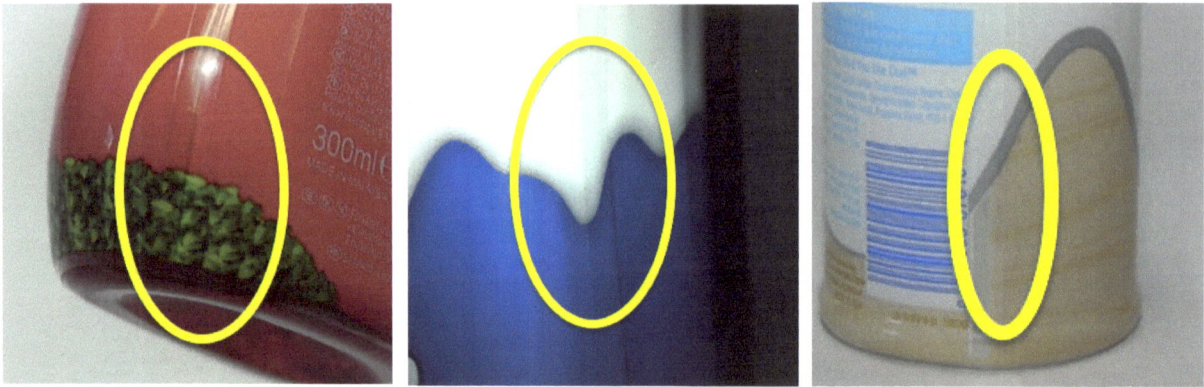

Figure 7.12, 7.13 and 7.14 Sleeved containers where the label designer achieved perfect graphic alignment in the seam area © 2017 Accraply, Inc

Figure 7.15 A shrink sleeved container with a horizontal barcode that is distorted © 2017 Accraply, Inc

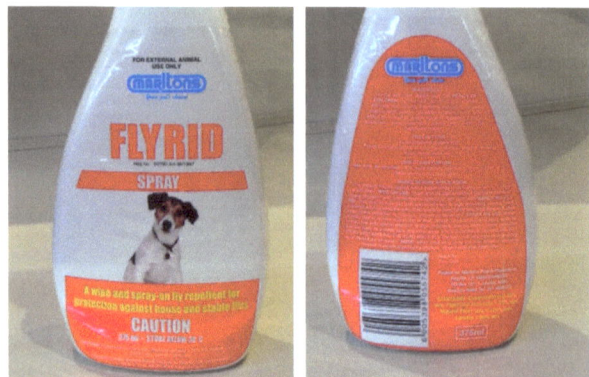

Figure 7.16 A shrink sleeved container with a proper vertical barcode © 2017 Accraply, Inc

THE IMPORTANCE OF INK SELECTION

Chapter 4 provided an extensive review of the unique properties required of shrink ink, as well as an overview of the shrink-specific considerations that come into play with each of the main print technologies.

The challenges faced by ink in the shrink sleeve world frequently do not become apparent until after the shrinking process, which underscores the need for testing – in the actual production tunnel, whenever possible.

Figure 7.18 illustrates the concept of 'ink pooling',

Figure 7.17 A shrink sleeve label that looks like a self-adhesive label © 2017 Accraply, Inc

Figure 7.18 Ink pooling causing a color change © 2017 Accraply, Inc

Figure 7.20 The container on the right shows evidence of 'tree barking' due to an excessive laydown of ink © 2017 Accraply, Inc

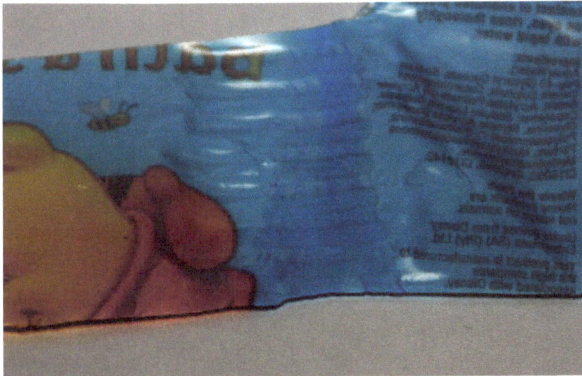

Figure 7.19 Removing the label shows the pooled ink and color change © 2017 Accraply, Inc

Figure 7.21 'Tree barking' from the inside of the sleeve © 2017 Accraply, Inc

in which a color change has occurred due to a thickening, or pooling, of the ink. The removed label in Figure 7.19 confirms the diagnosis. The causes of this defect are most likely due to fundamental problems with the quality of the ink, or the degree to which the ink has been formulated for the harsh demands of shrink sleeves. The problem may be compounded by a lack of sufficient 'slip characteristics' that allow for the ink and the sleeve to move freely over the ribbed surface of the container during the shrink process. It may also be compounded by the tunnel causing the sleeve to be

shrunk too quickly. A slower, more gradual shrink is usually more helpful.

Figure 7.20 illustrates the concept of 'tree barking' or 'orange peeling'. The container on the right is exhibiting the defect in the high shrink shoulder area, whereas the container on the left appears just fine. The likely difference is that the container on the left has a lower volume of ink in this area, meaning that there is less ink to 'bunch up' as the film shrinks. Figure 7.21 confirms that the ink chemistry has not managed to stay with the film in this high shrink area; there is a clear lack of adhesion in the high shrink area.

Figures 7.22, 7.23 and 7.24 each present the

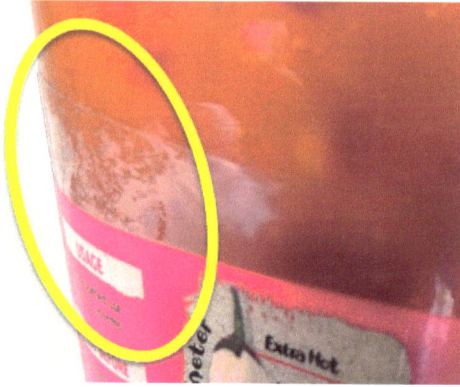

Figure 7.22 The 'Sellotape' effect on the transparent label area of this glass bottle © 2017 Accraply, Inc

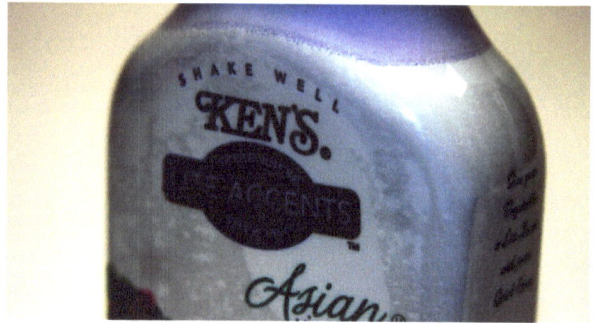

Figure 7.23 In this instance, the 'Sellotape' effect is not limited to just a transparent area of the label © 2017 Accraply, Inc

Figure 7.24 'The 'Sellotape' effect on a high shrink glass bottle © 2017 Accraply, Inc

Figure 7.25 ' The 'Sellotape' effect on a low shrink PET bottle © 2017 Accraply, Inc

most commonly misdiagnosed defect in shrink sleeve labeling. The question: is what we observe here is the 'wet T-shirt' effect or the 'Sellotape' effect? Chapter 4 alluded to these effects and the difference between them, despite that from a distance they present very similarly. In summary, the 'wet T-shirt' effect happens when moisture gets trapped between the sleeve and the container while in a steam tunnel. It will disappear in time as the moisture evaporates. The 'Sellotape' effect is permanent, however. It derives from how light is refracted as two smooth surfaces – the surface of the container and the inner surface of the film – come in close contact with each other. The effect is most noticeable in transparent areas (see Figure 7.22), but

as we can see in Figures 7.23, 7.24 and 7.25, it is not limited to just transparent areas. It is particularly evident from Figures 7.23 and 7.24 that the visual effect is most pronounced at those points where the film is under the most stress: on the corners of the containers. The effect is also more visible when the contents of the container are dark or black. This effect is often overcome by increasing the opacity of the ink when it is a printed sleeve, or by use of a clear varnish on clear sleeves or in transparent areas of printed sleeves. In any case, the most effective remedy is to introduce some separation between the two surfaces via a textured last down white or a varnish.

Figure 7.26 The raised edges on a slit roll © 2017 Accraply, Inc

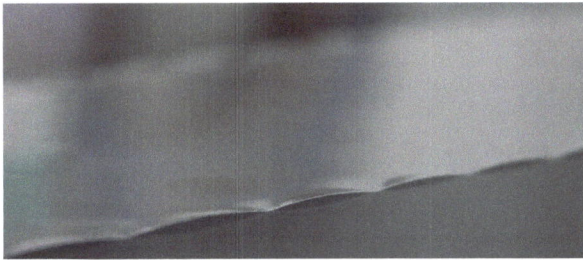

Figure 7.28 The flap that results from solvent not placed fully to the edge © 2017 Accraply, Inc

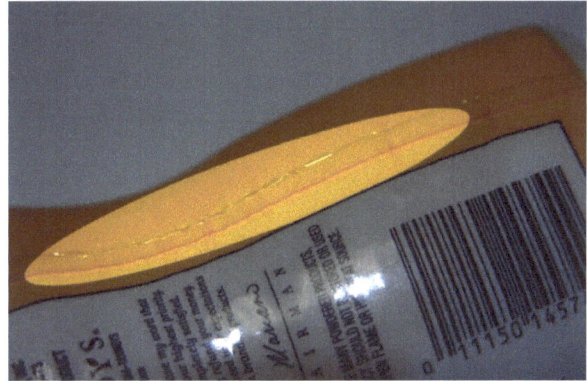

Figure 7.27 The scalloped slit edges on shrink film © 2017 Accraply, Inc

Figure 7.29 A seam where the solvent has not fully reached to the edge © 2017 Accraply, Inc

THE SECONDARY CONVERTING STEPS – SLITTING AND SEAMING

By way of recap on some of the main themes contained in Chapter 5, we will now draw attention to the three primary defects that emerge from the slitting and seaming steps in the process.

The recommendation to shear slit shrink film is clear from Chapter 5. Figure 7.26 illustrates the raised edges on a slit roll that can result from the use of razor slitting or from an incorrect implementation of shear slitting. Figure 7.27 illustrates the scalloped edges that result. These scalloped edges make it difficult, at best, to overlap those edges, lay down a bead of solvent, and create a perfect and mostly invisible seam. In fact, given that the solvent has the viscosity of water, an operator runs the risk of solvent flowing where it is not intended, which can cause blocking in the roll. This reality often forces seamer operators to pull back from placing solvent to the very edge – where it is required for the perfect seam – resulting in an unsightly seam, as shown in Figure 7.28.

Although the result, after shrinking, will not be as bad as that shown in Figure 7.28, the seamed sleeve in Figure 7.29 will also fall short of perfection. Why? Again, the answer lies in the slitting. In addition to shear slitting producing rolls of film without those scalloped edges, the correct implementation of shear slitting also produces a crisp and clean cut that facilitates the solvent flowing all the way to the edge. Without that crisp cut – or with the jagged cut that razor slitting delivers – the solvent does not make its way cleanly to the edge, and a

Figure 7.30 A burst seam © 2017 Accraply, Inc

Figure 7.32 A sleeve with cracked ink on the fold line © 2017 Accraply, Inc

Figure 7.31 A sleeve with a fold line © 2017 Accraply, Inc

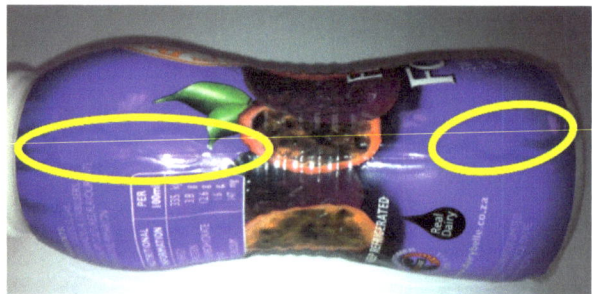

Figure 7.33 Fold lines that impeded the sleeve's ability to shrink smoothly © 2017 Accraply, Inc

less-than-perfect seam results. Figure 7.29 illustrates an example of where the solvent is not as 'to the edge' as perfection calls for.

Sticking with the theme of solvent, Figure 7.30 shows a burst seam. The burst has occurred in the area of greatest shrink where the seam is under most pressure. So, what may have caused the burst? Several explanations may exist, including: a solvent chemistry that is not a match for the film, resulting in a weak bond; contamination of ink or fountain solution in the seaming area, inhibiting a complete chemical bond; or a mechanical issue with the seamer that caused an insufficient delivery of solvent to this area.

A very common defect resulting from the converting step is the creation of creases along the fold lines of the 'tube' on the seaming equipment. Figure 7.31 shows unsightly fold lines that can be

traced back to the nip roll on the seamer. The solution here is to avoid crushing the edges as discussed in Chapter 5.

Figure 7.32 shows a starker example where the ink has actually cracked in the fold lines, allowing the white container to show through. Figure 7.33 is a different manifestation of the same fault where the impact of the fold lines has not cracked the ink, but it has impeded the ability of the film and the ink to shrink smoothly, and ink pooling has resulted.

SLEEVE APPLICATION AND SHRINKING
As we approach the finish line and set about getting sleeves applied to containers and then shrunk, there are numerous challenges that can emerge. The challenges with root causes that trace back to earlier steps in the process are difficult to remedy at this point, but those challenges that are specifically

Figure 7.34 Inconsistent sleeve application © 2017 Accraply, Inc

Figure 7.35 Inconsistent sleeve application © 2017 Accraply, Inc

Figure 7.36 Shrink result variation due to ink color © 2017 Accraply, Inc

associated with application and shrinking are certainly within our purview to correct.

With all shrink sleeve applications, there is a container handling component – that is, in the first instance, the containers should be delivered consistently to the applicator and the sleeves applied with consistent placement accuracy, relative to the shape of the container. Clearly, something went wrong in this regard with the products shown in Figure 7.34. The placement is different on each container, resulting in a very poor presentation on the retail shelf. Likely the shape of the container compounded the problem in this case, but this is an avoidable issue regardless. Moreover, the cut at the bottom and top of the sleeve is not in the right place, nor is it consistent. The same applies to Figure 7.35; using the heart-shaped icon on the lid as a center-line reference, we can observe just how inconsistently these labels were applied from one container to the next.

One of the earliest learnings newcomers to shrink sleeves make is the impact that the sleeve color can have on shrink results in a tunnel. Figure 7.36 shows identical glass bottles that were passed through the same tunnel – one with a black sleeve, and the other with a white sleeve. The white sleeve shrunk perfectly, whereas the black sleeve did not. Simple physics is in play – white reflects heat and black absorbs heat; hence, the same tunnel and the same settings deliver different results. The impact of color is generally more pronounced in a radiant or hot air tunnel than it is in a steam tunnel. A tunnel adjustment should solve the problem with the black sleeve, unless there are also issues with the ink or the volume of ink laydown in the high shrink area. Oftentimes, to get the desired color density with black, the volume of ink is increased to a point where it is difficult to achieve a good shrink result.

Remaining on the topic of shrinking, Figure 7.37 presents the problem known as 'fish eyes'. The

Figure 7.37 'Fish eyes' on a glass bottle © 2017 Accraply, Inc

Figure 7.38 'Fish eyes' and wrinkling on the shoulder and neck area of the bottle © 2017 Accraply, Inc

Figure 7.39 'Neck collapse' on high shrink containers © 2017 Accraply, Inc

Figure 7.40 'Neck collapse' on high shrink containers © 2017 Accraply, Inc

source of this problem was alluded to in Chapter 6 when glass was described as a heatsink. In other words, when the warm, and shrinking, film comes into contact with the colder surface of the glass bottle, the film's shrinking is stunted. Invariably, the solution is to pre-heat the glass container prior to applying the shrink sleeve label. Fish eyes tend to be a greater problem with radiant and hot air tunnels. Steam will generally produce a better result on cold containers, so long as there is enough dwell time in the tunnel. The result in Figure 7.37 may be further impacted by the seam limiting the shrink in this high shrink area.

Figure 7.38 presents similar characteristics with fish eyes and wrinkles presenting in the high-shrink shoulder area. The fish eyes are again the result of the container being a heatsink. While this container is not glass, perhaps it was a cold-filled product, which renders the same result. The wrinkles could be the result of the film not slipping sufficiently on the surface of the container. It is also likely that this product was shrunk in a radiant tunnel without the benefit of a spinning conveyor, which would have possibly facilitated a more even shrink result.

Figures 7.39, 7.40 and 7.41 all present identical issues around the neck of a bottle requiring a significant amount of shrink. In fact, this 'neck collapse' is a very common fault that is typically associated with attempts to shrink too quickly. As a general rule, you always want to shrink slowly, and from the bottom up. This progressive shrink may call for a zoned tunnel, or for multiple tunnels. In any case, the results shown in Figures 7.39, 7.40 and 7.41 are likely the result of tunnels that are too hot, causing the film to become soft and collapse on itself before the shrink to the bottle is complete.

Figure 7.42 presents an interesting opportunity to recap several learnings from throughout the book. The defect we are diagnosing is referred to as 'curl back', or 'flowering'. It is occurring because the sleeve was cut through the print – in other words, there is no clear area at the top of the sleeve. Those

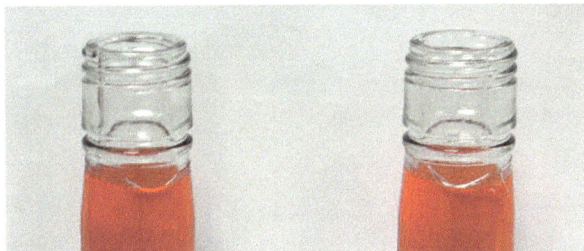

Figure 7.41 'Neck collapse' on high shrink containers © 2017 Accraply, Inc

Figure 7.42 'Curl back' issues around the neck of the bottle © 2017 Accraply, Inc

clear areas are important because, without the presence of ink, they tend to shrink faster and farther, and they serve to aggressively pull the shrinking sleeve tight at the top of the container. In this example, without this clear area to control the process at the top of the sleeve, the ink enters the shrink equation and creates the likelihood of a shrink differential between the inside and the outside of the film. In other words, the film wants to shrink faster than the ink, and the resulting curl always happens from the inside out, as demonstrated in Figure 7.42. It will be especially pronounced in areas of high shrink, where the density of ink is greatest.

In conclusion, we set out promising an exciting journey through the shrink sleeve label production process, with a view to achieving nothing short of perfection at the end. We have shown you the importance of selecting the right 'ingredients', partnering with your suppliers throughout, and focusing on the details in every step along the way. While we have illustrated the myriad challenges that exist throughout the process, we have also demonstrated the infinite opportunities presented by this exciting labeling technology. Despite the potential for mistakes to happen in the process of learning the science of shrink sleeve labeling, provided that appropriate care and consideration is given to every step of the process, the successes that result will drive your level of knowledge beyond that which is contained in this book. Every project presents its own learning opportunity and contribution to knowledge – your knowledge and that of the industry.

Index

www.ingramcontent.com/pod-product-compliance
Lightning Source LLC
Chambersburg PA
CBHW041721210326

41598CB00007B/733